The Evolution of Space-Time in the Early Universe

by

James Moffat

Foreword

This book is based on five peer reviewed journal papers which I published during the period 2017-18;

J Moffat, T Oniga et al (2017) 'Unitary Representations of the Translational Group Acting as Local Diffeomorphisms of Space-Time' J Phys Math 8:2, DOI: 10.4172/2090-0902.1000233

J Moffat, T Oniga et al (2017) 'A New Approach to the Quantisation of Paths in Space-Time' J Phys Math, 8:2, DOI: 10.4172/2090-0902.1000232. (2017).

J Moffat, T Oniga et al (2017) 'Ergodic Theory and the Structure of Non-Commutative Space-Time.' J Phys Math 8.2,. DOI:10.4172/2090-0902.1000229.

J Moffat, T Oniga et al, (2017) Factorial Unitary Representations of the Translational Group, Invariant Pure States and the Supersymmetric Graviton. J Phys Math 8: 257, DOI: 0.4172/2090-0902.1000257.

J Moffat C Wang (2018) Factorial Representations of Compact Lie Groups, Wigner Sets and Locally Invariant Quantum Fields. J Phys Math 9, 1, DOI: 10.4172/2090-0902.1000259

On reviewing and rereading these, it occurred to me that together they make an interesting mathematical story about the evolution of the universe.

This book is the result.

I hope you enjoy reading it as much as I enjoyed writing it.

I dedicate it to Paul Dirac, who was my inspiration.

Professor James Moffat

Dept of Physics, University of Aberdeen

2018

Contents

Chapter 1: Basic Notions

Quantum Theory is a guess at how inputs to an experiment at the quantum level relate to outputs. These inputs and outputs are real numbers read on displays in the domain of Newtonian mechanics. The experiment can be real or conjectural, leading to a test of validation or predictive power. Good guesses are those which are consistent with all the existing experimental data (the facts) and have high predictive power, within their domain of application. This approach will help us to address issues such as 'don't ask questions just calculate' as we will see. Our currently accepted theory is the simplest of the good guesses, because the simplest explanation consistent with the facts is, we assume, nearest to the truth.

From these preliminary remarks we can assume that a good model of quantum theory will have the following characteristics;

- ➤ It will have an elegant mathematical structure, since this is the simplest way of expressing its logic.
- ➤ It will be a theory based on the transformation of existing information into new information.
- ➤ It will be a black box model, with full access to its inputs and outputs but potentially incomplete information about why it works, since we do not require this of a good guess, model or theory. All we require is that for a given set of inputs, the outputs are correct, within its domain of application.

Wave Mechanics on Hilbert Space

For our purposes here, we define a free matter wave to be a disturbance which propagates with constant velocity in a given direction x. This disturbance is assumed to have a constant profile which does not decay over time. .It follows that the profile of the matter wave at time t, $\psi(x(t),t)$, is the same as the profile at time zero; $\psi(x(0),0)$, which equals $\psi(x(t)-ct,0)$ where c is the velocity with which the disturbance moves. Setting $\tau=x(t)-ct$ leads to the profile function $G(\tau) = \psi(\tau, 0)$ and the general wave equation ;

3

$$\frac{\partial^2 \psi(x(t),t)}{\partial t^2} = c^2 \frac{\partial^2 G(\tau)}{\partial \tau^2}; \frac{\partial^2 \psi(x(t),t)}{\partial x^2} = \frac{\partial^2 G(\tau)}{\partial \tau^2}.$$

$$\Rightarrow \frac{\partial^2 \psi(x(t),t)}{\partial x^2} = \frac{1}{c^2} \frac{\partial^2 \psi(x(t),t)}{\partial t^2}$$

A periodic plane wave solution has the typical form $\psi(x(t),t) = \exp 2\pi i\ (\frac{x}{\lambda} - \nu t)$

where ν is the frequency of this wave, λ the wavelength. Substituting the de Broglie relations for the energy E and momentum p of a free particle;

$$E = h\nu; \quad p = \frac{h}{\lambda}$$

into this wave amplitude function gives;

$$\psi(x,t) = \exp\frac{2\pi i}{h}(\frac{hx}{\lambda} - h\nu t) = \exp\frac{2\pi i}{h}(px - Et)$$

This is the most basic example possible of a wave function. It corresponds to a free running particle moving in the x direction with energy E and momentum p.

The set of all such wave functions for a single particle forms a space called a linear space or vector space. Given two wave functions $\psi_1(x,t); \psi_2(x,t)$ in this linear space we can also form the inner product;

$$\langle \psi_1(t) | \psi_2(t) \rangle = \int \psi_1^*(x,t)\psi_2(x,t)dx$$

The ability to form such inner products makes the linear space of wave functions into a Hilbert space. These wave functions relate to probabilities, as we will see later, so for a wave function state $\Psi(x,t)$ we also require that

$$\langle \psi(t) | \psi(t) \rangle = \int \psi^*(x,t)\psi(x,t)dx = 1$$

for all times t.

The Measurement Process

Now we describe the process of measuring this energy. Let's go back to our quantum wave function Ψ for a free particle moving in the direction x with energy E and momentum p. We first note that if we differentiate this expression in terms of time t then it pulls out the value E for the energy from the right-hand side, allowing it to be measured;

$$i\hbar \frac{\partial}{\partial t}\psi(x,t) = i\hbar \frac{\partial}{\partial t}\left\{ \exp \frac{2\pi i}{h}(px - Et) \right\}$$

$$= E \left\{ \exp \frac{2\pi i}{h}(px - Et) \right\}$$

$$= E\psi(x,t)$$

In other words, the linear operator $i\hbar \frac{\partial}{\partial t}$ has E as an eigenvalue for the energy eigenvector $\psi(x,t)$. Similarly, the linear operator $-i\hbar \frac{\partial}{\partial x}$ has momentum eigenvalues p;

$$-i\hbar \frac{\partial}{\partial x}\psi(x,t) = -i\hbar \frac{\partial}{\partial x}\left\{ \exp \frac{2\pi i}{h}(px - Et) \right\}$$

$$= p \left\{ \exp \frac{2\pi i}{h}(px - Et) \right\}$$

$$= p\psi(x,t)$$

5

Let $\hat{H} = i\hbar \dfrac{\partial}{\partial t}$ so that for our basic wavefunction ψ we have $\hat{H}\psi = E\psi$ .then;

➤ $\hat{H}$ is a self-adjoint linear operator since its eigenvalues are real numbers.

➤ Any reasonable wave function Ψ can be expressed as a weighted sum of these basic wavefunctions.

Taking these two together means that if Ψ_j are the basic wavefunctions corresponding to energy measurements E_j (such as the energy levels of an electron in a hydrogen atom), and ψ is a general wavefunction of the particle, then we can measure the energy of the quantum state defined by this general wavefunction as follows;

We have complex numbers α_j such that;

$$\psi = \sum_j \alpha_j \psi_j$$

And then the fact that the energy observable is a linear operator means that when we apply it to Ψ we get;

$$\hat{H}\psi = \sum_j \alpha_j \hat{H}\psi_j = \sum_j \alpha_j E_j \psi_j$$

Now instead of one energy level there is a weighted combination of several of them.

Because we insist that all wave function states, both the basic 'pure states' Ψ_j and the general one ψ are normalized, this implies that $\sum |\alpha_j|^2 = \langle \psi, \psi \rangle = 1$.

We interpret the $|\alpha_j|^2$ as probabilities adding up to one which means that they cover all the options.

Measurement of the energy of a general quantum state is then interpreted as follows.

With probability $|\alpha_j|^2$ we project down onto the basic eigenfunction Ψ_j and then the measured value of the energy is the corresponding eigenvalue E_j.

The Heisenberg – Dirac Viewpoint

From the point of view of linear operator theory (the Heisenberg – Dirac viewpoint), we can interpret this as follows.

> ➤ A quantum state is an element |x> of a Hilbert space which is either finite or separable.
> ➤ The inner product $< x, y >$ represents the (complex) amplitude of the transition from state $|x >$ to $|y >$.
> ➤ A system observable such as energy corresponds to a bounded linear operator acting on this state space.
> ➤ Energy measurements correspond to discrete eigenvalues of this linear operator.

If an ensemble of particles is measured, each prepared in the same initial energy state, then we will have a distribution of different energy measurements rather than a single value. Prediction of the nature of this distribution is a requirement of any theory. Only when the particle is in an eigenstate will the result be guaranteed. In quantum mechanics energy levels are spaced out or 'quantized' rather than smooth and continuous. We as humans don't see this discreteness directly because the spacing between energy levels is proportional to Planck's constant and this is a very small number. If we look at the energy emitted from a hydrogen atom through a spectrograph, however, we see these discrete levels. Each of these spectral lines corresponds to the energy emitted as a photon of light when the electron falls from a higher energy level (energy eigenstate and eigenvalue) to a lower one. If the levels were not quantized they would form a continuous blur.

The measured value (eigenvalue) E is the quantum version of kinetic energy if the particle is free running. However, we want to apply this process to particles like the electron trapped inside a Hydrogen atom and most certainly not free. Hydrogen is the simplest and lightest element consisting of a positively charged proton trapping a single negatively charged electron. The separation distance is about a tenth of a nanometer—at this small scale quantum effects are dominant.

From this 'picture' we can see that our electron has both kinetic and potential energy, the potential energy deriving from its location relative to the proton in the middle. This potential energy is due to the electrostatic attraction between particles of opposite charge (the + of the proton and the – of the electron) and the strength of the attraction goes as $\frac{1}{r^2}$ where r is the distance between the two particles—the smaller the distance the greater the force of attraction.

The Schrodinger Viewpoint

In the classical, non-quantum world, the sum of the kinetic and potential energies for a particle is called the Hamiltonian, denoted H. For our free particle we have just kinetic energy. The particle is not trapped. The classical (non-quantum) version of the Hamiltonian in this case is $H(x,t) = $ {kinetic energy at location x at time t}. Thus the total energy of the particle at location

x at time t =kinetic energy $E(x,t) = \frac{1}{2}mv^2 = \frac{(mv)^2}{2m} = \frac{1}{2m}p(x,t)^2$

Where the momentum is mv and v is the speed of the particle in the direction x. From this we have the classical Hamiltonian

$$H(x,t) = \frac{1}{2m}p(x,t)^2$$

We know that a quantum system has to look like the classical non-quantum system at a larger scale as we pull back from the detail. This is called the correspondence principle. We infer that the quantum version of the Hamiltonian must be of the form;

$$\hat{H}(x,t) = \frac{1}{2m}\hat{p}(x,t)^2$$

Substituting in the linear operators for total energy and momentum gives;

$$\hat{H} \rightarrow i\hbar\frac{\partial}{\partial t}; \hat{p} \rightarrow -i\hbar\frac{\partial}{\partial x}$$

$$\hat{H}\psi(x,t) = \frac{\hat{p}^2}{2m}\psi(x,t) \rightarrow i\hbar\frac{\partial}{\partial t}\psi(x,t) = -\hbar^2\frac{\partial^2}{\partial x^2}\psi(x,t)$$

This is Schrodinger's equation for a free particle. In the general case we also have a potential function of the form *V(x)*, normally depending only on position, as for the Hydrogen atom.

The Hamiltonian in this case is then the sum of the kinetic and potential energies;

$$\hat{H}(x,t)\psi(x,t) = \frac{1}{2m}\hat{p}(x,t)^2\psi(x,t) + V(x)\psi(x,t)$$

$$= -\frac{\hbar^2}{2m}\frac{\partial^2}{\partial x^2}\psi(x,t) + V(x)\psi(x,t)$$

Substituting for $\hat{H}$ gives the general Schrodinger equation

$$i\hbar\frac{\partial}{\partial t}\psi(x,t) = -\frac{\hbar^2}{2m}\frac{\partial^2}{\partial x^2}\psi(x,t) + V(x)\psi(x,t)$$

and the corresponding eigenvalue equation is $\hat{H}\psi = E\psi$ in other words;

$$-\frac{\hbar^2}{2m}\frac{\partial^2}{\partial x^2}\psi(x,t) + V(x)\psi(x,t) = E\psi(x,t)$$

To solve this equation, we assume that the location and time components of the wave function can be separated out. Thus this theory is inherently non-relativistic.

Relativistic waves

Maxwell's equations for the propagation of light waves through the vacuum of empty space can be written in the compact form;

$$\nabla \bullet E = 4\pi\rho; \ \nabla \bullet H = 0$$

$$\nabla \times E = -\frac{1}{c}\frac{\partial H}{\partial t}; \ \nabla \times H = \frac{1}{c}\frac{\partial E}{\partial t}$$

where ∇ is the gradient operator; *E* and *H* are now the standard terms for the electric and magnetic field strength vectors and ρ is a measure of the charge density. This symmetry between *E* and *H* leads to the plane wave solutions;

9

$$\nabla^2 E \equiv \frac{\partial^2 E}{\partial x^2} + \frac{\partial^2 E}{\partial y^2} + \frac{\partial^2 E}{\partial z^2} = \frac{1}{c^2} \frac{\partial^2 E}{\partial t^2} ; \ \nabla^2 H = \frac{1}{c^2} \frac{\partial^2 H}{\partial t^2}$$

If we consider the special case of a plane wave propagating along the x-axis, with the E vector pointing in the y-direction, we have;

$$E = (E_x, E_y, E_z) \text{ with } E_x = E_z = 0 \text{ and}$$

$$E_{y(t)} = E_{y(0)} \exp 2\pi i(vt - kx) = E_{y(0)} \exp 2\pi i v (t - \frac{x}{c})$$

We have from Maxwell's equations that:

$$\frac{\partial H_x}{\partial t} = -c\left(\nabla_y E_z - \nabla_z E_y\right)$$

$$\frac{\partial H_y}{\partial t} = -c\left(\nabla_z E_x - \nabla_x E_z\right)$$

$$\frac{\partial H_z}{\partial t} = -c\left(\nabla_x E_y - \nabla_y E_x\right)$$

Most of these expressions disappear, leaving just;

$$\frac{\partial H_z}{\partial t} = -c\nabla_x E_y = c\frac{2\pi i v}{c} E_{y(t)} = 2\pi i v E_{y(0)} \exp 2\pi i v (t - \frac{x}{c})$$

Thus the H vector has a non-zero z-component which varies with the same phase as the non-zero y-component of the E vector.

We now consider a plane periodic wave disturbance $F(x(t), y(t), z(t), t)$ propagating through free space in the direction given by the vector $\boldsymbol{r}$ with direction cosines (l, m, n). This means that

$\frac{r}{|r|}$ is the three dimensional vector $\left(le_x, me_y, ne_z\right)$ where e_x, e_y and e_z are unit vectors along the three space axes and $l^2 + m^2 + n^2 = 1$.

In three dimensions the set $\{x, y, z;\ lx + my + nz = |r|\}$ now defines a plane orthogonal to the direction of propagation r. The vector k defined by the components $\frac{v}{c}(l, m, n) = \frac{1}{\lambda}(l, m, n)$ is called the *wavevector*. For an arbitrary vector p drawn from the origin and ending in the point (x,y,z) of the plane we now have that $k.p = \frac{v}{c}(lx + my + nz)$. The equation of the disturbance can thus be written as;

$$F(p(t), t) = \exp 2\pi i (k.p \pm vt)$$

where p is an arbitrary 3 vector connecting the origin to a point on the plane orthogonal to the direction of propagation. The wavevector k points in the direction of propagation.

We now assume as a convention, that our wave is of fixed wavelength (i.e. it is monochromatic light) and propagates in the positive z-direction. This means that the E and H vectors are confined to the plane defined by the x and y axes, normal to the propagation direction.

A classical light wave which is polarized in the x-direction has a space-time dependent electric field oscillating in the x-direction which is the real part of the expression;

$E = E_0 e_x \exp 2\pi i (kz - vt)$.

In the same way, we can consider a y–polarized light wave which is the real part of;

$E = E_0 e_y \exp 2\pi i (kz - vt)$.

Circular polarization (i.e. circular rotation of the E vector in the $x - y$ plane) is produced by linearly combining an x-polarised plane E wave with a y-polarised plane E wave 90 degrees out of phase. The equation for such a wave is thus of the form;

$$E = E_0 \left(\frac{1}{\sqrt{2}} e_x \exp 2\pi i (kz - vt) + \left(\exp i \frac{\pi}{2} \right) \frac{1}{\sqrt{2}} e_y \exp 2\pi i (kz - vt) \right)$$

$$= E_0 \left(\frac{1}{\sqrt{2}} e_x \exp 2\pi i (kz - vt) + \frac{i}{\sqrt{2}} e_y \exp 2\pi i (kz - vt) \right)$$

These various classical linear combinations can more compactly be described by *Jones vectors,* which record the weights of the *x* and *y* components of *E*.

An *x*-direction polarized beam with weight $E_0 e_x$ thus corresponds to the Jones vector

$$\begin{pmatrix} E_0 e_x \\ 0 \end{pmatrix} = E_0 e_x \begin{pmatrix} 1 \\ 0 \end{pmatrix}$$ while a *y*-direction polarized beam corresponds to $\begin{pmatrix} 0 \\ E_0 e_y \end{pmatrix} = E_0 e_y \begin{pmatrix} 0 \\ 1 \end{pmatrix}$ and the

circularly polarized beam above corresponds to the Jones vector $\dfrac{1}{\sqrt{2}} \begin{pmatrix} E_0 e_x \\ i E_0 e_y \end{pmatrix}$. Normalizing these

so that the total weight in each case is 1 leads to the vectors $\begin{pmatrix} 1 \\ 0 \end{pmatrix}$, $\begin{pmatrix} 0 \\ 1 \end{pmatrix}$ and $\dfrac{1}{\sqrt{2}} \begin{pmatrix} 1 \\ i \end{pmatrix}$.

The Quantum Photon and Wigner theory

It is now possible to use the correspondence principle to argue that a polarized beam is composed of a stream of equivalently 'polarized' photon states defined by these normalized Jones vectors in a 2-dimensional vector space, giving rise to the well-known two-state qubit formulation of a photon. This is a good way of building an initial intuition. The basic quantum axioms tell us we need to define the 'observable' spin matrix *S* and its measured spin values correspond then to the eigenvalues of this matrix. How do we do this?

The answer was essentially supplied by Wigner who studied the group of rotations of space-time and their representation as groups of operators on vector spaces. It brings into play another property of light; its speed is independent of the source of the light; shining a torch out of a moving car does not make the photons go faster.

Consider now a massive particle of mass *m*, moving slowly with velocity $V \ll c$.

In two dimensions, let the vertical axis be time t and the horizontal axis be one-dimensional space *x*. The positive velocity *V* is represented by a vector pointing from the origin of these axes into the upper right quadrant to the point $(\Delta x, \Delta t)$ within the 'positive light cone'. Speed is distance over time (the vector pointing angle) so changes in the speed correspond to vector rotations in space-time. If c , the speed of light, is a finite universal constant then

$c^2 (\Delta t)^2 - (\Delta x)^2 = 0 = (i c \Delta t)^2 + (\Delta x)^2$ in any coordinate system. Moving a line around and

preserving its length s; where $s^2 = (i c \Delta t)^2 + (\Delta x)^2$ is Euclidean geometry in space-time. Thus the constancy of the speed of light is equivalent to using coordinates ict and x and assuming space-time is Euclidean ('flat').

In this Euclidean space-time a change of speed is still just a rotation, so we have the coordinate transformation;

$$\begin{pmatrix} \Delta x' \\ ic\Delta t' \end{pmatrix} = \begin{pmatrix} \cos\theta & -\sin\theta \\ \sin\theta & \cos\theta \end{pmatrix} \begin{pmatrix} \Delta x \\ ic\Delta t \end{pmatrix}$$

$$\Rightarrow \Delta x' = \cos\theta\, \Delta x - ic\sin\theta\, \Delta t = \cos\theta\left(\Delta x - ic\tan\theta\, \Delta t\right)$$

$$\Delta x' = 0 \Rightarrow \Delta x = ic\tan\theta\, \Delta t \Rightarrow \tan\theta = \frac{V}{ic}$$

$$1 = \cos^2\theta + \sin^2\theta \Rightarrow \frac{1}{\cos^2\theta} = 1 + \tan^2\theta = 1 - \frac{V^2}{c^2}$$

$$\Rightarrow \Delta x' = \cos\theta\left(\Delta x - ic\tan\theta\, \Delta t\right) = \frac{1}{\sqrt{1 - \frac{V^2}{c^2}}}\left(\Delta x - ic\left(\frac{V}{ic}\right)\Delta t\right) = \frac{1}{\sqrt{1 - \frac{V^2}{c^2}}}\left(\Delta x - V\Delta t\right)$$

Corresponding to the four vector *(cΔt, Δx, Δy, Δz)* in space-time, there is the energy-momentum vector; composed of four dimensionally consistent components $\left(\frac{E}{c}, p_x, p_y, p_z\right)$.

Multiplying the dimensionally consistent space-time vector *(cΔt, Δx, Δy, Δz)* by $\frac{m}{\Delta t}$ gives $\left(mc, p_x, p_y, p_z\right)$ and we equate this to the energy-momentum vector;

$$\left(mc, p_x, p_y, p_z\right) = \left(\frac{E}{c}, p_x, p_y, p_z\right)$$

Since the energy-momentum vector is unique, the equivalence of mass and energy comes from equating the first element of each expression.

Wigner Little Groups

Returning to the massless case, If a photon is travelling with momentum p along the z-axis, then its total energy is $e = pc$ and the energy-momentum vector is then $(p, 0, 0, p)$. This is invariant only to the 'Wigner little group' of rotations in the x-y plane orthogonal to the direction of propagation (z). Rotations operating on this plane have the form;

$$\hat{R}(\theta) = \begin{pmatrix} \cos\theta & \sin\theta \\ -\sin\theta & \cos\theta \end{pmatrix} \text{ with normalised eigenstates } \frac{1}{\sqrt{2}}\begin{pmatrix} 1 \\ i \end{pmatrix} \text{ and } \frac{1}{\sqrt{2}}\begin{pmatrix} i \\ 1 \end{pmatrix}$$

These eigenstates correspond to the Jones vectors for circular polarization of the classical EM wave, bringing the quantum and classical approaches into line. Relative to these eigenstates the spin operator $\hat{S}$ is defined as $\hat{R}(\theta) = \exp\left(\frac{i}{\hbar}\hat{S}(\theta)\right)$ with eigenvalues ± 1.

The Dirac Equation and Dirac Operator

The Dirac equation describes the dynamics of the quantum state of a spin ½ fermion (such as an electron) in a way which is consistent with the requirements of special relativity. This quantum state is four dimensional, with the first two components corresponding to the particle (e g the electron) and the second two components corresponding to the anti-particle (e g the positron). The Dirac spinor matrices γ^μ (μ=0,1,2,3) are themselves 4x4 matrices. They satisfy the following 'Clifford algebra' relationship;

$$\{\gamma^\mu, \gamma^\nu\} = \gamma^\mu\gamma^\nu + \gamma^\nu\gamma^\mu = \eta^{\mu\nu}I_{4x4}$$

where I_{4x4} is the 4x4 identity matrix (1s down the diagonal and 0 elsewhere) and $\eta^{\mu\nu}$ is the metric signature of flat space-time.

In essence, by defining the operator;

$$i\hat{D} = \sum_{j=0}^{3} \gamma_j \frac{\partial}{\partial x_j}; \quad \hat{D}^2 = \nabla^2$$

Dirac was able to take the square root of the relativistic relationship

$$E^2 = p^2c^2 + m^2c^4 \Leftrightarrow -\hbar^2\frac{\partial^2\psi}{\partial t^2} = \hbar^2\frac{\partial^2\psi}{\partial x^2} + m^2c^4\psi \Leftrightarrow \hat{D}^2\psi = m^2c^4\psi \text{ with } p \text{ the momentum of the}$$

particle. In quantum operator terms, this results in both positive and negative square roots, corresponding to the particle and its antiparticle. These Dirac spinor matrices automatically incorporate the spin of the particle, showing that spin is an inherently relativistic effect. In more detail, we take here the 'Weyl representation' of the Dirac matrices so that;

$$\gamma^\mu = \begin{pmatrix} 0 & \sigma^\mu \\ \bar{\sigma}^\mu & 0 \end{pmatrix}$$

14

Where σ^μ (μ=0,1,2,3) are the 2x2 Pauli spin matrices, $\bar{\sigma}^0 = \sigma^0$ and $\bar{\sigma}^\mu = -\sigma^\mu$ (μ=1,2,3).

Given these relationships we can show directly that $i\gamma^0\gamma^1\gamma^2\gamma^3 = \begin{pmatrix} -I_{2x2} & 0 \\ 0 & I_{2x2} \end{pmatrix}$

where I_{2x2} is the 2x2 identity matrix $\begin{pmatrix} 1 & 0 \\ 0 & 1 \end{pmatrix}$.

This operator representation acts on a 4 dimensional vector space (in fact a 4 dimensional Hilbert space). Each normalized element of this vector space is a quantum state ψ.

Defining γ^5 to be the composite matrix operator $i\gamma^0\gamma^1\gamma^2\gamma^3$, we can write this state ψ in the form

$\psi = \frac{1}{2}(I_{4x4} - \gamma^5)\psi + \frac{1}{2}(I_{4x4} + \gamma^5)\psi$. We define these two terms as the left landed and right

handed spinor components of the state, thus we define;

$\psi_L = \frac{1}{2}(I_{4x4} - \gamma^5)\psi$ and $\psi_R = \frac{1}{2}(I_{4x4} + \gamma^5)\psi$

So that $\psi = \psi_L + \psi_R$.

Now from above we know that $(I_{4x4} - \gamma^5) = \begin{pmatrix} I_{2x2} & 0 \\ 0 & I_{2x2} \end{pmatrix} - \begin{pmatrix} -I_{2x2} & 0 \\ 0 & I_{2x2} \end{pmatrix} = \begin{pmatrix} 2I_{2x2} & 0 \\ 0 & 0 \end{pmatrix}$

Similarly, $I_{4x4} + \gamma^5 = \begin{pmatrix} 0 & 0 \\ 0 & 2I_{2x2} \end{pmatrix}$.

Thus the four vector ψ_L has only two non-zero elements (the first two) and is of the form

$\psi_L = \begin{pmatrix} \psi_1 \\ \psi_2 \\ 0 \\ 0 \end{pmatrix}$. Similarly, the four vector ψ_R has only two non-zero elements (the last two) and is

defined as $\psi_R = \begin{pmatrix} 0 \\ 0 \\ \bar{\chi}^{\dot{1}} \\ \bar{\chi}^{\dot{2}} \end{pmatrix}$. These then correspond to the left handed 'chiral' Weyl spinor $\begin{pmatrix} \psi_1 \\ \psi_2 \end{pmatrix}$ and

the right handed chiral Weyl spinor $\begin{pmatrix} \bar{\chi}^{\dot{1}} \\ \bar{\chi}^{\dot{2}} \end{pmatrix}$. In four dimensions we have $\psi = \begin{pmatrix} \psi_L \\ \psi_R \end{pmatrix}$ which is an

element of a four dimensional Hilbert space. The left handed Weyl spinor quantum state ψ_L

corresponds to the particle and the right handed Weyl spinor quantum state ψ_R corresponds to the anti-particle.

If ψ is a general Dirac spinor, it is a solution of the massless form of the Dirac equation:

$$\sum_{\mu} i\gamma^{\mu}(\partial_{\mu} - ieA_{\mu})\psi = 0$$

where ψ has charge e and A_{μ} ($\mu = 0,1,2,3$) is the vector potential corresponding to an external magnetic field. Note that the case with mass m non-zero simply introduces an additional term into the Dirac equation of the form $m\psi$ and the same argument goes through.

Chapter 2: The Evolution of Space-Time in the Early Universe

In perturbative forms of Quantum Gravity, it is typically assumed that a background spacetime exists. A stage is provided, and the role of physics is to correctly model the dynamical interactions of the actors on that stage. However, a fundamental theory ideally should model the evolution of the stage as well as the players. The early universe was photon rich, but the mean free path of such early photons was constrained by the extremely dense, hot nature of the material created following the quantum fluctuation of the big bang. Only when the universe became transparent would the photon velocity in empty space become a universal constant. In terms of the early evolution of space-time the key relativistic property was spin. In my own work, discrete steps across a network of entangled spins form the basis of space-time and are the foundation on which more elaborate constructs can be established.

There is still no settled quantum theory of gravity, thus what is described below represents only one way of potentially creating such a fundamental theory. In constructing this I have been greatly influenced by the views of Abhay Ashtekar.

Spinors and Spin Networks

As described in Chapter 1, a left-handed Weyl 2-spinor is an element of a 2-dimensional vector space F with a basis of Clifford variables denoted ψ_A (A = 1,2).

Given such a 2-spinor ψ_A, the 2x2 matrix M acting on ψ_A gives rise to another 2-spinor $\psi'_A = M_A^B \psi_B$, summing over repeated indices.

Left-handed and right-handed Weyl spinors are related as follows;

$$\psi_A \in F \text{ implies } \psi_A^* = \bar{\psi}_{\dot{A}} \in \dot{F}$$

The representations $M \to M$ and $M \to \left(M^{-1}\right)^T$ have the same dimensionality and are therefore unitarily equivalent; there is a unitary matrix $\varepsilon = \varepsilon_{AB}$ such that for all M,

$$\varepsilon^{-1} M \varepsilon = \left(M^{-1}\right)^T \text{ where } \varepsilon = \begin{pmatrix} 0 & -1 \\ 1 & 0 \end{pmatrix}.$$

Defining ε^{AB} by $\left(\varepsilon^{AB}\right)^{-1} = \begin{pmatrix} 0 & -1 \\ 1 & 0 \end{pmatrix}$ we have $\varepsilon^{AB} = \begin{pmatrix} 0 & 1 \\ -1 & 0 \end{pmatrix}$.

The matrices ε_{AB} and ε^{AB} are tensor objects which can be used to raise and lower indices in the usual vector and tensor calculus, within which left-handed 'chiral' Weyl 2-spinors are covariant vectors, and right-handed chiral Weyl 2-spinors are contravariant vectors. Denoting the dual space as F^*. the following calculation illustrates this 'spinor calculus' for $\psi \in F^*$ and $\chi \in F$;

$$\psi\chi = \psi^A \chi_A = \varepsilon^{AB}\psi_B \chi_A = \varepsilon^{12}\psi_2\chi_1 + \varepsilon^{21}\psi_1\chi_2 = \psi_2\chi_1 - \psi_1\chi_2$$

We also define the matrix $\bar{\varepsilon}$ such that the representations

$M \to M^*$ and $M \to \left(M^{*-1}\right)^T$ are equivalent, so that $\bar{\varepsilon}^{-1} M^* \bar{\varepsilon} = \left(M^{*-1}\right)^T$ for all M^*.

We now consider in this Weyl 2 - spinor context the Dirac 4x4 spin matrix representation;

$\gamma^\mu = \begin{pmatrix} 0 & \sigma^\mu \\ \bar{\sigma}^\mu & 0 \end{pmatrix}$ where σ^μ (μ=0,1,2,3) are the Pauli 2 x 2 spin matrices.

In this representation, we can show directly that $i\gamma^0\gamma^1\gamma^2\gamma^3 = \begin{pmatrix} -I_{2x2} & 0 \\ 0 & I_{2x2} \end{pmatrix}$ where I_{2x2} is the 2x2

identity matrix $\begin{pmatrix} 1 & 0 \\ 0 & 1 \end{pmatrix}$.

This Dirac spin operator representation acts on a split 4-dimensional vector space. Each element of this vector space suitably normalized, is a quantum ket state $|\psi>$; a pure state. Defining γ^5 to be the composite matrix operator $i\gamma^0\gamma^1\gamma^2\gamma^3$, we can write this state $|\psi>$ in the

split form $|\psi> = \frac{1}{2}(I_{4x4} - \gamma^5)|\psi> + \frac{1}{2}(I_{4x4} + \gamma^5)|\psi>$. These correspond to the left landed and

right handed 2-spinor components of the state;

$$\psi_L = \frac{1}{2}(I_{4x4} - \gamma^5)\psi \text{ and } \psi_R = \frac{1}{2}(I_{4x4} + \gamma^5)\psi$$

So that $\psi = \psi_L + \psi_R$.

Thus, the four vector ψ_L has only two non-zero elements (the first two) and is of the

form $\psi_L = \begin{pmatrix} \psi_1 \\ \psi_2 \\ 0 \\ 0 \end{pmatrix}$. Similarly, the four vector ψ_R has only two non-zero elements (the last

two) and is defined as $\psi_R = \begin{pmatrix} 0 \\ 0 \\ \overline{\chi}^{\dot{1}} \\ \overline{\chi}^{\dot{2}} \end{pmatrix}$.

Assuming planar isotopy, it is possible to associate various locally deformable lines in the plane with Weyl 2-spinor calculations, giving rise to topological structures called spin networks. As a

simple example the tensors $\overline{\varepsilon} = \varepsilon^{\dot{A}\dot{B}} = \begin{pmatrix} 0 & 1 \\ -1 & 0 \end{pmatrix}$ and $\overline{\varepsilon}^{-1} = \varepsilon_{\dot{A}\dot{B}} = \begin{pmatrix} 0 & -1 \\ 1 & 0 \end{pmatrix}$ correspond to;

Their product is then a closed loop which corresponds to a scalar λ times the identity matrix with $\lambda = 1$ in this case. The advantage of this approach is that spinor calculations become a sequence of potentially simpler topological transformations with connections to knot theory[1].

Computational Spin Networks

Building on these ideas, we define a *computational spin network* to be a finite quiver consisting of a directed graph with n nodes, where the nodes represent entangled spin inputs and a directed link between two nodes corresponds to a quantum gate, as we now discuss.

If we represent the basis spin $\pm\frac{1}{2}$ eigenvectors as the column vectors $0 >= \begin{pmatrix} 1 \\ 0 \end{pmatrix}$ and $1 >= \begin{pmatrix} 0 \\ 1 \end{pmatrix}$,

then the NOT quantum gate which switches $|0>$ to $|1>$ and $|1>$ to $|0>$ corresponds to the

unitary matrix $U = \begin{pmatrix} 0 & 1 \\ 1 & 0 \end{pmatrix}$. It is easy to check that $U|0>=|1>$ and $U|1>=|0>$. All quantum

gates correspond in this way to multiplying the input state vector by a unitary matrix.

---

[11] Two 3-dimensional knots are equivalent if there is a sequence of such two-dimensional transformations linking one knot to the other (the 'Reidemeister moves').

To ease notational clutter, in all that follows we denote the joint tensor product state of the spins |x> and |y> as |xy>.

Given the entangled state, $|\beta(0,0)\rangle = \frac{1}{\sqrt{2}}(|00\rangle + |11\rangle)$ and $|y_0\rangle = \frac{1}{\sqrt{2}}(\alpha|0\rangle + \beta|1\rangle)$;

Let $|y_1\rangle = |y_0\rangle \otimes |\beta(0,0)\rangle = \frac{1}{2}\alpha|0\rangle \otimes |\beta(0,0)\rangle + \beta|1\rangle \otimes |\beta(0,0)\rangle$

$= \frac{1}{2}(\alpha|0\rangle \otimes (|00\rangle + |11\rangle) + (\beta|1\rangle \otimes |00\rangle + |11\rangle)$

$\equiv \frac{1}{2}(\alpha|000\rangle + \alpha|011\rangle) + (\beta|100\rangle + \beta|111\rangle)$

Define $|y_2\rangle = U(CNOT)|y_1\rangle$ mapping;

$|00\delta\rangle \rightarrow |00\delta\rangle$; $|01\delta\rangle \rightarrow |01\delta\rangle$; $|10\delta\rangle \rightarrow |11\delta\rangle$ and $|11\delta\rangle \rightarrow |10\delta\rangle$ where $\delta \in \{0,1\}$

$\Rightarrow |y_2\rangle = \frac{1}{2}(\alpha|000\rangle + \alpha|011\rangle) + (\beta|110\rangle + \beta|101\rangle)$

Replacing $|y_0\rangle = \frac{1}{\sqrt{2}}(\alpha|0\rangle + \beta|1\rangle)$ by using the Hadamard unitary gate $U(H)$ we have

$U(H)|y_0\rangle = \frac{1}{\sqrt{2}}\alpha U(H)|0\rangle + \beta U(H)|1\rangle = \frac{1}{2}\{(\alpha|0\rangle + \alpha|1\rangle) + (\beta|0\rangle - \beta|1\rangle)\}$

Then we have;

$|y_2\rangle = \frac{1}{2}(\alpha|000\rangle + \alpha|011\rangle) + (\beta|110\rangle + \beta|101\rangle)$

$= \frac{1}{2}((\alpha|0\rangle(|00\rangle + |11\rangle) + \beta|1\rangle(|10\rangle + |01\rangle))$

$\Rightarrow |y_3\rangle = \frac{1}{2}\{\alpha U(H)|0\rangle(|00\rangle + |11\rangle) + \beta U(H)|1\rangle(|10\rangle + |01\rangle)\}$

$= \frac{1}{2}\{(\alpha|0\rangle + \alpha|1\rangle)(|00\rangle + |11\rangle) + (\beta|0\rangle - \beta|1\rangle)(|10\rangle + |01\rangle)\}$

$= \frac{1}{2}\{\alpha|000\rangle + \alpha|011\rangle + \alpha|100\rangle + \alpha|111\rangle + \beta|010\rangle + \beta|001\rangle - \beta|110\rangle - \beta|101\rangle\}$

$= \frac{1}{2}\{|00\rangle(\alpha|0\rangle + \beta|1\rangle) + |01\rangle(\alpha|1\rangle + \beta|0\rangle) + |10\rangle(\alpha|0\rangle - \beta|1\rangle) + |11\rangle(\alpha|1\rangle - \beta|0\rangle)\}$

This can then be interpreted in terms of the original vector $|y_0\rangle$ as;

$$\frac{1}{2}\{|00>U(00)|y_0> + |01>U(01)|y_0> + |10>U(10)|y_0> + |11>U(11)|y_0>\}$$

Comparing the two expressions implies that;

$$U(00) = I_{2x2} = \begin{pmatrix} 1 & 0 \\ 0 & 1 \end{pmatrix}; U(01) = \begin{pmatrix} 0 & 1 \\ 1 & 0 \end{pmatrix}^{-1}; U(10) = \begin{pmatrix} 1 & 0 \\ 0 & -1 \end{pmatrix}^{-1}; U(11) = \begin{pmatrix} 0 & 1 \\ -1 & 0 \end{pmatrix}^{-1}$$

These are all real unitary matrices thus their inverse is simply the transpose matrix in each case;

$$U(00) = I_{2x2} = \begin{pmatrix} 1 & 0 \\ 0 & 1 \end{pmatrix}; U(01) = \begin{pmatrix} 0 & 1 \\ 1 & 0 \end{pmatrix}; U(10) = \begin{pmatrix} 1 & 0 \\ 0 & -1 \end{pmatrix}; U(11) = \begin{pmatrix} 0 & -1 \\ 1 & 0 \end{pmatrix}$$

Thus, we can transport a Quantum State along such a computational path.

Note that $U(00) = \sigma^0$ $U(01) = \sigma^1$ $U(10) = \sigma^3$ $iU(11) = \sigma^2$ where $\{\sigma^j; j = 0,1,2,3\}$ are the Pauli spin matrices.

We identify this transport from node to node of the computational spin network with the action of a sequence of discrete translations in space-time from one node to the next neighboring node. A discrete path in space-time can be then be considered as a series of applications of the translation subgroup of the Poincare group.

We next show there is a local mapping from this translation group into a neighbourhood of the identity of a quantum Weyl algebra fibre bundle, such that the whole classical path can be lifted into the fibre bundle to form a unique quantum field as a section through the fibres. Under the further assumptions of a scale invariant relativity, we can also show that a discrete closed loop in space time, corresponding to two classical paths sharing the same end points, is renormalisable and the finite limit is then a fractal path in smooth, macroscopic space-time.

Quantum Paths in Space-Time

We start by considering classical phase space. Given a dynamical system, entropy is defined through considering the phase space of the system. The emergent behavior of this classical system gives rise to regions of phase space, each corresponding to similar macro-level behaviour. The number and variation in size of these regions reflects the overall complexity of the system. The entropy of such a coarse-grained region is essentially a count of all the different micro-configurations constituting that region. A system starting in a low entropy state will tend to wander into larger coarse-grained volumes; hence thermodynamic entropy tends to increase over time if the system is isolated, giving rise to the second law of thermodynamics. The structure of classical phase space is such that each set of initial conditions (x_μ, p_μ) generates a unique solution; solutions do not cross. For a Hamiltonian system it is thus possible to reformulate classical mechanics as a symplectic vector space of solutions, or an equivalent set of initial conditions of location and momentum, equipped with a bilinear form Ω which ultimately derives from Hamilton's equations of motion;

With $\xi \equiv (x, p)$ a 6-dimensional vector we have $\dfrac{\partial \xi}{\partial t} = \Omega \dfrac{\partial H}{\partial \xi}$ where $\Omega = \begin{bmatrix} 0_{3\times3} & I_{3\times3} \\ -I_{3\times3} & 0_{3\times3} \end{bmatrix}$.

This is of the form of a symplectic vector space $V \oplus V^*$ where V is a real finite vector space with dual V^*. In a way similar to Hodge's reformulation of Maxwell's equations, the skew-symmetric rank 2 tensor Ω then takes the general form;

$$\Omega(x \oplus \eta, x' \oplus \eta') = \eta' \cdot x - \eta \cdot x'.$$

In our case V is the configuration space, V^* the (dual) momentum space and $V \oplus V^*$ the phase space, a product vector bundle over V with fibre V^*. By choosing values such as $(1,0,0,0,0,0)$ we can pull out elements;

$$\Omega(x \oplus \eta, 0 \oplus \eta') = \eta' \cdot x = \eta'_1.$$

From this point of view the Dirac canonical quantisation of elements of phase space such as η'_1 is equivalent to the canonical quantisation; $\Omega \to \hat\Omega$ as a (not necessarily bounded) linear operator, and this form of canonical quantisation extends smoothly to countably infinite phase space [1].

Given this canonical quantisation; $\Omega \to \hat\Omega$, we can form the Weyl unitaries $\hat{W} = \exp i\hat\Omega$. Closure of linear combinations of these unitaries and their adjoints in the normed operator topology forms the Weyl algebra.

The extension of Ω to the space of solutions allows us to define an inner product on S as $\langle S(1), S(2) \rangle = -i\Omega\big(S(1)^*, S(2)\big)$ where $S(1)^*$ is the complex conjugate solution. It turns out [1] that this defines an inner product on S relative to which a one particle Hilbert space can be defined. For a quantum system of bosonic harmonic oscillators, we can then assemble a symmetric tensor product Fock space in the usual way, using creation and annihilation operators.

An example of the existence of the Weyl form in a two dimensional locally flat space-time is now given for a local algebra $O(D)$, having a representation as observables acting on the Hilbert space $L^2(x,t)$ with Lebesgue measure.

For a small increment of space-time $(\delta x, \delta t)$ we consider the Poincare Translation subgroup element $T(\delta x, \delta t):(x,t) \rightarrow (x+\delta x, t+\delta t)$ and define;

$$U_{T(\delta x, \delta t)} f(x,t) = f(x-\delta x, t-\delta t) \text{ for } f(x,t) \in L^2(x,t)$$

Then U is a local group homomorphism of the translation group T as observables acting on $L^2(x,t)$. Define also;

$$V_{\delta t} = U_{T(0,\delta t)}; V_{\delta x} = U_{T(\delta x,0)}$$
$$\Rightarrow V_{\delta t} V_{\delta t'} f(x,t) = U_{T(0,\delta t)} U_{T(0,\delta t')} f(x,t) = f(x, t-\delta t'-\delta t)$$
$$= U_{T(0,\delta t'+\delta t)} f(x,t) = V_{\delta t+\delta t'} f(x,t) \Rightarrow V_{\delta t} V_{\delta t'} = V_{\delta t+\delta t'}$$

A similar result applies for $V_{\delta x}$, by symmetry.

Now introduce a deformation of the form; $T(\delta x, 0) \rightarrow Z_{\delta x} f(x,t) = \exp(it\delta x) f(x-\delta x, t)$.

The mappings V and Z are unitary representations on $L^2(x,t)$ and so also is their product

$$V \times W : (\delta x, \delta t) \rightarrow T(\delta x, \delta t) \rightarrow V_{\delta t} Z_{\delta x}$$

We have;

$$V_{\delta t} Z_{\delta x} f(x,t) = V_{\delta t} \exp(it\delta x) f(x-\delta x, t)$$
$$= \exp(it\delta x) f(x-\delta x, t-\delta t)$$
$$Z_{\delta x} V_{\delta t} f(x,t) = Z_{\delta x} f(x, t-\delta t)$$
$$= \exp(i(t-\delta t)\delta x) f(x-\delta x, t-\delta t)$$
$$= \exp(-i\delta t\delta x) V_{\delta t} Z_{\delta x} f(x,t)$$
$$Z_{\delta x} V_{\delta t} = \exp(-i\delta t\delta x) V_{\delta t} Z_{\delta x}$$

Then $T(\delta x, \delta t) \rightarrow (Z_{\delta x}, V_{\delta t})$ is a local Weyl representation of the CCR on $L^2(x,t)$. By the Stone-von Neumann theorem, the resulting C*-algebra and its weak closure as a von Neumann algebra must be unitarily isomorphic to Wald's equivalent 'algebraic approach' to quantum field construction and his Weyl Algebra, since we can assume all relevant Hilbert spaces are separable in application to observed real systems.

This example indicates that a continuous local group homomorphism from a neighbourhood of the identity of T to a neighbourhood of the identity of the set of observables in O(D) exists as a Weyl algebra. It can be easily extended to 4 dimensions by replacing x by the 3-vector $\mathbf{x} = (x_1, x_2, x_3)$.

A discrete path in space-time, as we have so far generated it, can be considered as a set of linked causally directed intervals each of fractal dimension 1 in renormalized smooth space-time. More formally, we define the path as a series of n linked increments $\mathbf{a}(j)$ with varying direction relative to a local forward light cone, such that the path begins at $\mathbf{x(0)}$ and ends at $\mathbf{x(1)}$, with $T(\mathbf{a}(j)): \mathbf{x} \to \mathbf{x} + \mathbf{a}(j)$ elements of the translation subgroup T. The total path is then generated by the product

$\prod_{j=1}^{j=n} T(\mathbf{a}(j)) \mathbf{x(0)}$ with the final end point $\mathbf{x(1)} = \mathbf{x(0)} + \sum_{j=1}^{j=n} \mathbf{a}(j)$. For a fixed initial point $\mathbf{x(0)}$ we

can identify this path with the finite group product $\prod_{j=1}^{j=n} T(\mathbf{a}(j)) \in T$.

Now let there be given a continuous local group homomorphism from a neighbourhood V of the identity of T to the neighbourhood W of the identity of the set of observables in $O(D)$ as a Weyl representation of the CCR. Then a discrete classical path CP in space-time can be lifted to a quantum field section QP through $O(D)$.

Proof. We construct the section iteratively, following a method suggested by [2]. Let CP be a discrete classical path in space-time. From the definition, the path is a series of n linked increments $\mathbf{a}(j)$ each of Euclidean interval length $|\mathbf{a}(j)|$ and varying direction such that the path begins at $\mathbf{x(0)}$ and ends at $\mathbf{x(1)}$, with $T(\mathbf{a}(j)): \mathbf{x} \to \mathbf{x} + \mathbf{a}(j)$ elements of the translation subgroup T.

We have, by assumption, a continuous local group homomorphism $\varphi: T(a) \to U_{T(a)}$ from V, a neighbourhood of the identity of T to W, a neighbourhood of the identity of $(O(D))$. Clearly, by redefining the number of links in CP if necessary, we can assume that $|\mathbf{a}(j)|$ is sufficiently small so that $T(\mathbf{a}(j)) \in V$ for all j and by choosing appropriate units we can assume that CP consist of n links each of length $1/n$.

We also assume that for n large; $|\mathbf{x}(t_2) - \mathbf{x}(t_1)| < \dfrac{1}{n} \Rightarrow T(\mathbf{x}(t_1))^{-1} T(\mathbf{x}(t_2)) = T(\mathbf{a}(t_1)) \in V$.

Let m be a positive integer strictly less than n and suppose that the path QP has been defined such that its initial value is $QP(0) = A(\mathbf{x}(\mathbf{0}))$. We proceed by induction. Assume that QP has been defined for all values of $\mathbf{x}(t)$ with $0 \le t \le \dfrac{m}{n}$, and satisfies, for all such t with $0 \le t \le \dfrac{m}{n}$;

(a). the fixed endpoint assumption; $QP(0) = A(\mathbf{x}(\mathbf{0}))$;

(b). the local lifting assumption to the Weyl algebra near the identity; If $\mathbf{x}(s)$, $\mathbf{x}(t)$ in CP satisfy $|\mathbf{x}(s) - \mathbf{x}(t)| \le \dfrac{1}{n}$ then

$T(\mathbf{x}(s))^{-1} T(\mathbf{x}(t)) \in V$ and $\varphi\left(T(\mathbf{x}(s))^{-1} T(\mathbf{x}(t))\right) = QP(\mathbf{x}(s))^{-1} QP(\mathbf{x}(t)) \in W$

We now extend the path $\{QP(\mathbf{x}(t)); \ 0 \le t \le \dfrac{m}{n}\}$, stepping forward one additional link on

CP so that $t = \dfrac{m}{n} + \dfrac{1}{n}$; using the following construction;

$$QP\left(\mathbf{x}\left(\frac{m+1}{n}\right)\right) = QP\left(\mathbf{x}\left(\frac{m}{n}\right)\right) \varphi\left(T\mathbf{x}\left(\frac{m}{n}\right)^{-1} T\mathbf{x}\left(\frac{m+1}{n}\right)\right) \quad \ldots\ldots\ldots(1)$$

From equation (1) the extension of the path QP still satisfies (a): $QP(0) = A(\mathbf{x}(0))$ since φ acting on the identity of the group local translations T is the identity operator in O(D). We need to show that the extension under induction still satisfies the local lifting assumption (b).

Let h be a real number with $|h| \le \dfrac{1}{n}$. If h is positive, then by induction h satisfies the extension shown in equation (1). Thus, we have;

$$QP\left(\mathbf{x}\left(\frac{m}{n} + h\right)\right) = QP\left(\mathbf{x}\left(\frac{m}{n}\right)\right) \varphi\left(T\mathbf{x}\left(\frac{m}{n}\right)^{-1} T\mathbf{x}\left(\frac{m}{n} + h\right)\right)$$

25

If on the other hand h is negative then setting $\mathbf{x}(s) = \mathbf{x}\left(\dfrac{m}{n} + h\right)$ and $\mathbf{x}(t) = \mathbf{x}\left(\dfrac{m}{n}\right)$; since s,

t is both equal to or less than $\dfrac{m}{n}$ by the inductive hypothesis they therefore satisfy;

$$\varphi\left((T\mathbf{x}(s))^{-1} T\mathbf{x}(t)\right) = (QP\mathbf{x}(s))^{-1} QP\mathbf{x}(t) \text{ with } t = \frac{m}{n} + h \text{ and } s = \frac{m}{n}$$

$$\Rightarrow \varphi\left(\left(T\mathbf{x}\left(\frac{m}{n}\right)\right)^{-1} T\mathbf{x}\left(\frac{m}{n} + h\right)\right) = \left(QP\mathbf{x}\left(\frac{m}{n}\right)\right)^{-1} QP\mathbf{x}\left(\frac{m}{n} + h\right)$$

$$\Rightarrow QP\mathbf{x}\left(\frac{m}{n} + h\right) = QP\mathbf{x}\left(\frac{m}{n}\right) \varphi\left(T\mathbf{x}\left(\frac{m}{n}\right)^{-1} T\mathbf{x}\left(\frac{m}{n} + h\right)\right)$$

Thus equation (1) holds for both positive and negative values of h.

It follows that;

$$\left(QP\mathbf{x}\left(\frac{m+1}{n}\right)\right)^{-1} QP\mathbf{x}\left(\frac{m}{n} + h\right) = \left(QP\mathbf{x}\left(\frac{m}{n}\right) \varphi\left(T\mathbf{x}\left(\frac{m}{n}\right)^{-1} T\mathbf{x}\left(\frac{m+1}{n}\right)\right)\right)^{-1} \left(QP\mathbf{x}\left(\frac{m}{n}\right) \varphi\left(T\mathbf{x}\left(\frac{m}{n}\right)^{-1} T\mathbf{x}\left(\frac{m}{n} + h\right)\right)\right)$$

$$= \varphi\left(T\mathbf{x}\left(\frac{m+1}{n}\right)^{-1} T\mathbf{x}\left(\frac{m}{n} + h\right)\right)$$

Thus, the path extension satisfies both requirements (a) and (b) completing the inductive

step $\dfrac{m}{n} \to \dfrac{m+1}{n}$, provided we satisfy the local topological constraints, namely;

We have that if n is sufficiently large then there is a neighbourhood U such that;

$$|x(t_2) - x(t_1)| < \frac{1}{n} \Rightarrow \text{ by continuity } T(x(t_1))^{-1} T(x(t_2)) = T(a(t_1)) \in U^{-1}U \subset V$$

$$\text{and } \varphi\left(T(x(t_1))^{-1} T(x(t_2))\right) = \varphi(T(a(t_1))) \in W$$

Setting, for small $h > 0$;

$$t_1 = \frac{m+1}{n}; t_2 = \frac{m}{n} + h \Rightarrow |\mathbf{x}(t_1) - \mathbf{x}(t_2)| \text{ small}$$

$$\Rightarrow T\mathbf{x}\left(\frac{m+1}{n}\right)^{-1} T\mathbf{x}\left(\frac{m}{n} + h\right) = T\left(\mathbf{x}(t_1)\right)^{-1} T\left(\mathbf{x}(t_2)\right) = T\left(\mathbf{x}(t_2)\right) T\left(\mathbf{x}(t_1)\right)^{-1} = T\left(\mathbf{x}(t_2) - \mathbf{x}(t_1)\right) \in V$$

$$\varphi\left(T\left(\mathbf{x}\left(\frac{m+1}{n}\right)\right)^{-1} T\mathbf{x}\left(\frac{m}{n} + h\right)\right) \in W \text{ as required.}$$

26

Since $\varphi(\text{identity of } T) = \text{identity of } O(D)$ the induction hypothesis is true for $\frac{m}{n} = 0; h = \frac{1}{n}$

. By induction the path QP(t) can be extended in O(D) for all discrete steps m less than or equal to n.

We now show that the constructed quantum field QP is unique.

Proof. The initial point of QP is unique by condition (a).

If $QP(\mathbf{x}(t))$ is unique for all $t \leq t_0$ then let $t_0 < t \leq t_0 + \varepsilon$. Then;

$$T\mathbf{x}(t_0)^{-1}T\mathbf{x}(t) \in V \Rightarrow \varphi\left(T\mathbf{x}(t_0)^{-1}T\mathbf{x}(t)\right) = QP\mathbf{x}(t_0)^{-1}QP\mathbf{x}(t) \Rightarrow QP\mathbf{x}(t) = QP\mathbf{x}(t_0)\varphi\left(T\mathbf{x}(t_0)^{-1}T\mathbf{x}(t)\right)$$

Thus, the path QP is uniquely determined for all points $t < t_0 + \varepsilon$. The result follows by induction.

Finally, we show that there is a projection π from the fibre bundle $O(D)$ mapping the quantum field back to the path CP.

Proof. From condition (a) this is clear for the initial point. We again use induction to prove the general case. If the projection π maps the observable '$QP\,\mathbf{x}(t)$' back to $T\mathbf{x}(t)$ for $t \leq t_0$, let $t_0 < t \leq t_0 + \varepsilon$. Then as before we have;

$$QP\mathbf{x}(t) = QP\mathbf{x}(t_0)\varphi\left(T\left(\mathbf{x}(t_0)\right)^{-1}T\mathbf{x}(t)\right)$$
$$\text{thus } \pi QP\mathbf{x}(t) = \pi QP\mathbf{x}(t_0)\pi \circ \varphi\left(T\left(\mathbf{x}(t_0)\right)^{-1}T\mathbf{x}(t)\right) = T\mathbf{x}(t_0)\left(T\left(\mathbf{x}(t_0)\right)^{-1}T\mathbf{x}(t)\right) = T\mathbf{x}(t)$$

The result follows by induction.

Renormalisation of Discrete Paths in Space-Time

The principle of relativity is captured within the assumptions of the Riemannian geometry of 4-manifolds, where formulae equating a tensor expression to zero remain invariant under local diffeomorphism transformations. It is a natural extension of these ideas to additionally postulate that the scales of measurement inscribed on the clocks or measuring rods used by an observer should also not be absolute. Mathematically this can be captured by the additional requirement that the tensor formulae should be invariant under transformations of scale. From this perspective a relativistic quantum system is a *scale free system*.

If Φ is the function transforming system inputs to system outputs, there are only two non-trivial possibilities available for this system under renormalisation of one of the similarity parameters;

a). Φ tends to a non-zero finite limit as $\Pi_2 \to 0$. This means that Φ can be replaced by its limiting expression, with complete separation of variables and the functional relationship is a product of powers whose values can be determined by dimensional analysis.

b). Φ has power law asymptotics of the form $\Phi = \Pi_2^{\alpha_1}\Phi'\left(\dfrac{\Pi_1}{\Pi_2^{\alpha_2}}\right)$, as $\Pi_2 \to 0$. The power law form of the limiting expression still leads to separation of variables, but with characteristic exponents equal to the 'anomalous' fractional dimensions of a Gell-Mann and Low form of renormalisation.

Under the assumptions of scale relativity, let us consider a discrete closed loop in space-time; corresponding to two discrete non-oriented paths sharing the same end points. Then this loop is renormalisable and has a finite limit corresponding to the curve fractal dimension.

Proof. Assume that we have a fractal closed loop L in space-time with Euclidean diameter d. We approximate L by a discrete closed path $L(\eta)$ where η is the Euclidean length of each segment of $L(\eta)$. Standard dimensional analysis shows that N(η), the number of segments in the path L(η), is a function of the form $f\left(\dfrac{d}{\eta}\right)$. We will establish the nature of this function and its renormalisation limit.

The fractal, self-similar nature of the discrete path implies that if we consider a finer segmentation of segment length ξ, then $N(\xi) \propto N(\eta)N(\xi\,|\,\eta) = \dfrac{f\left(\dfrac{d}{\eta}\right)f\left(\dfrac{\eta}{\xi}\right)}{f(1)^2}$ where $N(\xi\,|\,\eta)$ is the number of segments of length ξ in a segment of length η, and $N(d) = N(\eta\,|\,\eta) = f(1)$.

It follows that $\dfrac{f\left(\dfrac{d}{\xi}\right)}{f(1)} = \dfrac{f\left(\dfrac{d}{\eta}\right)f\left(\dfrac{\eta}{\xi}\right)}{f(1)^2}$ thus $f\left(\dfrac{d}{\xi}\right) = \dfrac{f\left(\dfrac{d}{\eta}\right)f\left(\dfrac{\eta}{\xi}\right)}{f(1)}$

This implies that f, for the limiting case, must be of the form $f\left(\dfrac{x}{y}\right) = C\left(\dfrac{x}{y}\right)^{D}$ with C

and D constants; $C = f(1)$. Thus, we have;

$$f\left(\frac{d}{\eta}\right) = f(1)\left(\frac{d}{\eta}\right)^{D} \Rightarrow N(\eta) = \left(\frac{d}{\eta}\right)^{D} \text{ and } L(\eta) = \eta\left(\frac{d}{\eta}\right)^{D}$$

Locally along the limiting smooth form this implies $L(\xi \mid \eta) = \xi\left(\dfrac{\eta}{\xi}\right)^{D} = \eta^{D}\xi^{1-D}$.

The renormalisation limit is thus finite and we identify D with the path fractal dimension.

We now investigate in more depth the subgroup T of the Poincare group consisting of translations of space-time as a gauge group of automorphisms. We define a representation of T as a group of automorphisms of a local fibre algebra $\mathbf{A(x)}$ which we assume to be isomorphic to a von Neumann algebra with trivial centre acting on a separable Hilbert space, rooted at the event point x.

Consider then the subgroup T acting on $\mathbf{A(x)}$. These actions generate the local diffeomorphisms of General Relativity. If we make the minimal assumption that this representation is weakly measurable; i.e. the mapping $g \rightarrow \langle \alpha_g (A)x, y \rangle$: is Haar-measurable for all relevant values of A, x and y; then the argument of [3] shows that the mapping $g \rightarrow \alpha_g$ is norm continuous. Since the translational group is both abelian and connected, it follows that each α_g is an inner automorphism of $\mathbf{A}$ and the corresponding unitary W_g has a spectrum contained in the positive half plane. In fact, we have;

$$\sigma(W_g) \subset \{z; \operatorname{Re} z \geq \beta_g\} \text{ where } \beta_g = \frac{1}{2}\left(4 - \| \alpha_g - i \|^2\right)^{\frac{1}{2}}$$

Moreover, if $\mathbf{S}$ denotes the von Neumann subalgebra generated by $\{W_g; g \in T\}$, the set of unitaries $\{W_g; g \in T\}$ is a commuting set within $\mathbf{A(x)}$. Thus, $\mathbf{S}$ is a commutative subalgebra and contains the identity I of $\mathbf{A(x)}$, since if id is the group identity then $I = W_{id}$. $\mathbf{A(x)}$ is a factor, $\mathbf{S}$ thus contains the centre Z of $\mathbf{A(x)}$. Such a commutative quantum operator algebra is equivalent to the set of continuous functions on a compact space and this equivalence arises through the Gelfand transform; $A \rightarrow \hat{A}$; $\hat{A}(\rho) = \rho(A)$. with ρ an element of the carrier space Φ_S of $\mathbf{S}$; the set of all such continuous complex valued homomorphisms. Φ_S is topologically a Stonean space, as is Φ_Z and the restriction map $\pi : \Phi_S \rightarrow \Phi_Z$ is a continuous surjection. We then have, as we will prove, a lifting

$$f : \Phi_Z \rightarrow \Phi_S \text{ with } f(\rho)|_z = \pi \circ f(\rho) = \rho \text{ for } \rho \in \Phi_Z.$$

For each $g \in T$, we can now define the Gelfand transform of a unitary U_g acting on the carrier space Φ_S of **S** as follows;

$$\hat{U}_g(\rho) = \overline{\hat{W}}_g(f(\rho|_z))\hat{W}_g(\rho) = \overline{\hat{W}}_g(f \circ \pi(\rho)\hat{W}_g(\rho), \quad \rho \in \Phi_S$$

If we define the equivalence $\rho \approx \rho' \Leftrightarrow \rho|_z = \rho'|_z$ then by the extended form of the Stone-Weierstrass theorem [4] the centre Z corresponds to those elements of Φ_S constant on each equivalence class, Applying this to the Gelfand transform $\hat{V}_g(\rho) = \overline{\hat{W}}_g(f(\rho|_z))$ it follows that $\hat{V}_g$ corresponds to an element V_g of the centre. Since **A(x)** is a factor, this means that $V_g = v(g)I$ with $v(g)$ a complex number defining a coboundary, and that for each $g \in T$; U_g implements α_g. In addition, the mapping $g \to U_g$ is a group homomorphism for if we set;

$R_{g,h} = U_g U_h U_{gh}^*$ then for an operator A, $R_{g,h} A R_{g,h}^* = U_g U_h U_{gh}^* A U_{gh} U_h^* U_g^* = \alpha_g \alpha_h \alpha_{gh}^{-1}(A) = A$
Hence $R_{g,h}$ is a unitary in the centre. In fact $R_{g,h} = \lambda(g,h)I$ where $\lambda(g,h)$ is a 2-cocycle.

The fact that the lifting $f : \Phi_Z \to \Phi_S$ has the property $\pi \circ f(\rho) = f(\rho)|_z = \rho$ for $\rho \in \Phi_Z$ means that $\hat{R}_{g,h}(\rho) = \hat{R}_{g,h}(f(\rho))$ since the domain of $\hat{R}_{g,h}$ is Φ_Z . But then we have;

$\hat{U}_g(f(\rho)) = 1 \quad \forall g \in G$, thus $\hat{R}_{g,h}(\rho) = 1 \quad \forall \rho \in \Phi_Z$; the 2-cocycle is trivial.
Therefore $R_{g,h} = I \quad \forall g,h \in G$ and $g \to U_g$ is a group representation by unitaries
 in the fibre algebra, which turns out to be norm continous, as previously shown [1, 7].

In summary, the lifting follows from the Axiom of Choice applied to the strange topological properties of Stonean spaces. We can set it in the context of lifting from a totally disconnected, compact Hausdorff base space B into a containing fibre bundle K having a Stonean topology.

Projection onto the Base Space B of a Stonean Fibre Bundle K.

We prove first the fact that the projection π is an open mapping if and only if for each non-trivial open subset E of K, $\pi(E)$ is not a nowhere dense subset of the base space B.
One way is trivial for if π is an open mapping then $\pi(E)$ is a non-trivial open set so cannot be nowhere dense.
Conversely, the topology of the fibre bundle K is compact and totally disconnected, with a basis of 'clopen' sets (i.e. sets which are both closed and open). Thus, every open set is a union of such clopen sets, and it suffices to show that if V is clopen in K, then $\pi(V)$ is open in B.

31

Since V is clopen, it is a closed and thus compact subset of K, and π (V) is continuous, thus π (V) is compact. If we define $Y = \pi(V) \setminus \text{int } \pi(V)$; this a closed set with empty interior thus Y is a nowhere dense set and is the image of an open set; $Y = \pi\left(V \setminus \pi^{-1}(\text{int } \pi(V))\right)$, using the fact that the projection mapping π is continuous and surjective. It follows that Y is void and $\pi(V) = int$ $\pi(V)$. Thus $\pi(V)$ is open and π is an open mapping.

A Unique quantum field isomorphic to the Base Space

Consider now the set Φ of all compact subsets S of K such that π (S) = B. Then Φ is a non-void, partially ordered by set inclusion, and every decreasing chain has a lower bound. It follows from Zorn's Lemma that Φ has a smallest element K(0). We show that;

> $\pi|_{K(0)}$ is an open mapping ;

> $\pi|_{K(0)}$ is bijective

This will prove the result; the Axiom of Choice selecting out K(0) as a unique minimal section through the fibre bundle K.

$\pi|_{K(0)}$ is an open mapping ;

Consider then V to be a non-trivial open set in K(0). By definition, $\pi|_{K(0)}$ is surjective, thus;

$B \setminus \pi(V) \subset \pi(K(0) \setminus V)$. Now $K(0) \setminus V$ is a closed thus compact subset of K, and

$\pi|_{K(0)}$ is continuous, thus $\pi(K(0) \setminus V)$ is compact. It follows that $(B \setminus \pi(V))^{-} \subset \pi(K(0) \setminus V)$. If $\pi(V)$ is nowhere dense, then;

$$B = B \setminus \text{int}(\pi(V))^{-} \subset B \setminus \pi(V)^{-} \subset \pi(K(0) \setminus V), \text{ a closed, compact set.}$$

This is a contradiction due to minimality of the set K(0). Thus $\pi(V)$ cannot be a nowhere dense set. From our earlier discussion, this is enough to show that $\pi|_{K(0)}$ is an open mapping.

$\pi|_{K(0)}$ is injective ;

Assume that $\pi|_{K(0)}$ is not injective, then $\exists x_1, x_2; x_1 \neq x_2 \in K(0)$ with $\pi(x_1) = \pi(x_2)$.

Since K(0) is a Hausdorff topological space with a basis of clopen sets, there is a clopen subset V of K(0) containing x_1 but not x_2. Then;

$\pi(V)$ is also clopen and $\pi^{-1}(\pi(V)) \setminus V \subset K(0)$, and $\pi(x_2) = \pi(x_1) \in \pi(V)$, thus $x_2 \in \pi^{-1}(\pi(V)) \setminus V$.

It follows that $W = \pi^{-1}(\pi(V)) \setminus V$ is a non-trivial open set, and W and V are disjoint, with

$B \setminus \pi(V) \subset \pi(K(0) \setminus W)$ and $\pi(V) \subset \pi(K(0) \setminus W) \Rightarrow \pi(K(0) \setminus W) = B$

This again contradicts the minimality of K(0). Thus $\pi|_{K(0)}$ is a continuous bijection, and the required lifting is given by $f = \left(\pi|_{K(0)}\right)^{-1}$.

Mackey Theory and Fibre Bundle Liftings on Stonean Base Space

A character acting on a locally compact abelian group G is a continuous group homomorphism from G to the unit circle. If H is a closed subgroup of G and ρ is a character of H, then ρ extends to a character on G.

Since this useful result plays an important part in our argument, we provide a simple proof as follows, based on Mackey's general approach.

Let $\hat{G}$ denote the group of characters on G under pointwise multiplication. Pontryagin [2] shows that we can impose a locally compact topology on $\hat{G}$ relative to which it is a locally compact abelian group. Define;

$$\omega(g)(\rho) = \rho(g) \quad \forall g \in G, \rho \in \hat{G}$$

Then the mapping ω is a continuous isomorphism between G and $\hat{\hat{G}}$.

Let $L = \left\{\rho \in \hat{G}; \rho(H) = 1\right\}$ and $S = $ the character group of the quotient group $\dfrac{G}{H}$

then $L = S$ for if $\beta \in S \Leftrightarrow \alpha: g \to \beta(Hg) \Leftrightarrow \alpha(h) = 1 \quad \forall h \in H \Leftrightarrow \alpha \in L$

If we move this up a level and now define S as the character group of the quotient group $\dfrac{\hat{G}}{L}$

then S is the subset of $G = \hat{\hat{G}}$ which is constant on L but up to isomorphism, $H = \hat{\hat{H}}$ is constant on L thus in this case $S = H$.

Finally, we can conclude that, up to isomorphism, $\hat{H} = \hat{S} = \dfrac{\hat{G}}{L}$ since S is the character group of

the quotient group $\dfrac{\hat{G}}{L}$.

Let φ denote the isomorphism: $\hat{H} \to \dfrac{\hat{G}}{L}$. If ψ denotes the restriction mapping

$\rho \to \rho|_H : \hat{G} \to \hat{H}$, then the kernel of ψ is L. Thus $\psi = \varphi^{-1}$. If Ω is the quotient mapping:

$\hat{G} \to \dfrac{\hat{G}}{L}$ then we have;

For $\xi \in \hat{H}$ the mapping $\Omega^{-1} \circ \varphi : \hat{H} \to \dfrac{\hat{G}}{L} \to \hat{G}$ thus $\rho = \Omega^{-1} \circ \varphi(\xi) \in \hat{G}$

and then $\rho|_{\hat{H}} \equiv \psi(\rho) \equiv \psi(\rho L) \equiv \psi \circ \Omega(\rho) = \xi$ since $\varphi^{-1} = \psi$

and $\xi \to \rho$ is the require extension to the character group G of the character group of the subgroup H.

By exploiting Mackey Theory directly, we now aim to prove our previous key result by an alternative method linking to the short exact sequences of group cohomology theory.

Define $\alpha : g \to \alpha_g$ to be a representation of the translational subgroup T of the Poincare group as a gauge group of automorphisms of the fibre algebra A(x). Provided this group representation α is weakly measurable, then it turns out that it is also norm, weakly and strongly continuous, and is implemented by a norm, weakly and strongly continuous unitary representation by unitaries in the fibre algebra.

Proof. With these assumptions, the mapping $\alpha : g \to \alpha_g$ is norm continuous and is thus implemented by a set $g \to W_g$ of commuting unitaries in the fibre algebra [3].

Since $\| \alpha_g - i \| \geq \{ \| \alpha_g(T)x - Tx \|; \ \ \| T \| \leq 1 \| x \| \leq 1 \}$ it follows that the norm continuous mapping $\alpha : g \to \alpha_g$ is strongly continuous. Our foundational work at [5] then shows that we may choose the unitaries in a way that the mapping $g \to W_g$ is a borel mapping from T to the unitary group of **A(x)** with the weak operator topology.

Let $N = Ker(\alpha) = \{ g \in T ; \alpha_g = \text{the identity} \}$. Replacing T by T/N we may assume that;

$$\alpha_g = \alpha_h \text{ implies } g = h. \ldots\ldots\ldots(1).$$

We note that the unitary group of the centre of **A(x)** is isomorphic to the unit circle **O** in the complex plane. It is also easy to see that $\Gamma = \{ \lambda W_g ; \lambda \in \mathbf{O}, \text{ and } g \in T \}$ is an abelian subgroup of the unitary group of **A(x).**

Define the mapping $\gamma(\lambda) = \lambda I$ which maps the unit circle **O** into the subgroup Γ, and the mapping $\eta(\lambda W_g) = g$ which maps the group Γ to the translation group T. Note that the mapping η is well defined here due to (1) above. Then the sequence;

$$0 \to \mathbf{O} \xrightarrow{\gamma} \Gamma \xrightarrow{\eta} T \to 0$$

is short exact. Thus, Γ is an extension of $\mathbf{O}$ by T. We can identify the group Γ with the extension $\mathbf{T}\eta'T$ where; $\eta'(\lambda, g) = \eta(\lambda W_g) = g$ via the mapping; $(\lambda, g) \to \lambda W_g$.

Let the mapping J denote the identification $\mathbf{O} \times T \leftrightarrow \Gamma \leftrightarrow \mathbf{O}\eta'T$ then J is a borel mapping. Thus, we see again that the mapping $W : g \to W_g$ is a borel measurable mapping from the translation subgroup T into the group Γ.

The group T is a separable locally compact abelian group and is a standard borel space. The unitary group of $\mathbf{A(x)}$ is also a standard borel space, thus $W^{-1} : W_g \to g$ is a borel mapping and η' is a borel system of factors for $\mathbf{T}\eta'T$. There is thus a locally compact topology on the group Γ relative to which it is a separable, locally compact abelian group. Γ also contains the unit circle $\mathbf{O}$ as a closed subgroup.

With this Mackey topology on Γ, we can now deduce that the identity map $e^{i\theta} \to e^{i\theta} : \mathbf{O} \to \mathbf{O}$ is a character on $\mathbf{O}$ and thus lifts to a character $\sigma : \Gamma \to \mathbf{O}$.

For each unitary $W_g; g \in T$ define $U_g = \sigma(W_g)^* W_g$. Then;

U_g implements the automorphism τ_g and, since $\sigma|_o$ is the identity;

$$\sigma(U_g) = \sigma(W_g)^* \sigma(W_g) = 1 \, \forall g \in T.$$

Now if :

$$g, h \in T \text{ then } U_g U_h = \lambda(g,h) U_{gh} \text{ for a 2-cocycle } \lambda(g,h) \in \mathbf{O}$$

Since $U_g \in \Gamma \, \forall g$, by definition of Γ, we can apply the group homomorphism σ to both sides of this relationship. Noting again that $\sigma|_o$ is the identity, this implies that all such 2-cocycles are trivial, and the mapping $g \to U_g$ is a unitary group representation.

From the proof above we can identify the group Γ with the borel system of factors $\mathbf{O}\eta'T$

The mapping $W : g \to W_g$ is a borel measurable mapping from the translation subgroup T into the group Γ. Hence the mapping $g \to \langle x, W_g x \rangle$ is a measurable mapping for any x in the Hilbert space H. In addition, the character $\sigma : \Gamma \to \mathbf{O}$ is a continuous mapping, thus $\sigma \circ W : g \to \sigma(W_g)$ is borel. Combining these together it follows that the mapping

$g \to \langle x, U_g x \rangle = \sigma(W_g)^* \langle x, W_g x \rangle$ is borel measurable and is in fact norm continuous due as before to the spectral properties of the

$\{W_g; g \in G\}$ which imply that $\|U_g - I\| \le 2\sqrt{2 - 2\beta_g}$ and $\beta_g \to 1$ as $g \to e$.

$\{W_g; g \in G\}$ which imply that $\|U_g - I\| \le 2\sqrt{2 - 2\beta_g}$ and $\beta_g \to 1$ as $g \to e$.

The Poincare group is a locally compact Lie group with 10 generators, and the translational group is an abelian subgroup generated by the energy-momentum 4-vector P_μ. This has the property that its square $P^2 = P_\mu P^\mu = E^2$ lies in the centre of the Lie algebra. If we consider the energy-momentum vector in normalized units *(c=1)* then P^2 has the form $P^2 = m^2 I$, where m is the mass-energy of the corresponding particle. In other words, a factorial representation of the Translational group corresponds to a particle with fixed mass m and undetermined spin. We can consider two cases;

(a). $P^2 = m^2.I; m^2 \neq 0$ corresponding to a multiplet of particles each of the same positive mass but with different spin values;

(b). $P^2 = 0.I$. This factorial representation corresponds to a massless particle such as a photon or a graviton, with a Supersymmetric massless fermion partner.

We continue to assume as context that space-time is non-commutative at some energy level such as the Planck regime, with algebraic structure at each event point **x** of space-time, forming the fibre algebra **A(x)**. This structure then corresponds to the single fibre of a principal fibre bundle. A gauge group of automorphisms corresponding to the translation subgroup T of the Poincare group acts on each fibre algebra locally, while a section through this bundle is then a quantum field of the form $\{A(x); x \in M\}$ with M the underlying space-time manifold. In addition, we assume a local algebra $O(D)$ corresponding to the algebra of sections of such a principal fibre bundle with base space a finite and bounded subset of space-time, D. The algebraic operations of addition and multiplication are assumed defined fibrewise for this algebra of sections.

We define a *separating family* of T-invariant quantum states to be a finite or countable subset S of the state space such that the positive kernel of S is zero; i.e. given an observable A in the positive subset of the fibre algebra **A(x)** ,

$$f_j(A) = 0 \ \forall f_j \in S \Rightarrow A = 0$$

For a general observable A we assume that $f(A^*A) = 0 \Rightarrow A = 0$. This reflects the classical case; if a tensor or the difference between two tensors of the same covariant and contravariant class is equal to zero for all local inertial reference frames, then it is zero for all curvilinear reference frames, by the covariance assumptions of General Relativity.

Since the Hilbert space F(x) on which the fibre algebra **A(x)** acts is separable, by definition, there is a countable dense subset $x(n)$. Defining $f = \sum_1^\infty \alpha_{x(n)} \omega_{x(n)}; \lim_{n \to \infty} \sum_1^n \alpha_{x(j)} = 1$; clearly f is a separating state, and then each element of the weak*-closed convex hull $\Delta = \overline{co}\{f \bullet \alpha_g ; g \in T\}$ is also separating, for if;

$$\sum_j \lambda(j) f \circ \alpha(g_j)(A^*A) = 0 \Rightarrow \exists g_j \in T; f \circ \alpha(g_j)(A^*A) = 0 \Rightarrow \alpha(g_j)(A^*A) = 0,$$

since f is separating.

Applying the automorphism $\alpha(g_j^{-1}) \Rightarrow A*A = 0$. Additionally, if $f_n(A*A) \to g(A*A)$ and $f_n(A*A) = 0$ then $g(A*A) = 0$. Thus every element of Δ is separating.

It follows, by applying the Hajian-Kakutani fixed point theorem [6] to the weak*- compact set Δ that it contains an invariant state. Thus, there is a separating T-invariant quantum state. This has potential structural implications for the Wigner-Dirac representation theory of the underlying algebra A(x).

If f is a separating T-invariant quantum state, let π be the GNS representation then clearly π is an algebraic isomorphism since the kernel of f is $\{0\}$. Let $\{U_g ; g \in G\}$ be the Segal unitaries associated with f [7] then the mapping $g \to U_g$ is a unitary representation implementing $\alpha : g \to \alpha_g$. If the mapping $\alpha : g \to \alpha_g$ is weakly measurable in the GNS representation then by our previous results [3, 8] it is norm continuous and we can further assume that there is a unitary mapping $g \to U_g$ implementing $\alpha : g \to \alpha_g$ which is norm continuous with

$$U_g \in \pi(\mathbf{A(x)}) \ \forall g \in T \ .$$

We can characterise these invariant states in terms of their ergodic action focusing on density matrix states and their von Neumann entropy. From inspection, we can generalise many of these previous results to any (not necessarily normal) quantum state as discussed later in an appendix.

We have, from that analysis; that the automorphic representation $\alpha : g \to \alpha_g$ of T acts ergodically if and only if $\pi(A(x)) \cap \{U_g ; g \in T\}'$ is trivial, containing only the projections 0 and I and thus consisting of the set of complex multiples of I.

We also have;

$$J\left\{\pi(A(x)) \cap \{U_g ; g \in T\}'\right\} J = \pi(A(x))' \cap \{U_g ; g \in T\}' \ .$$

Since $J^2 = 1$, it follows that the representation $\alpha : g \to \alpha_g$ of T acts ergodically if and only if

$\pi(A(x))' \cap \{U_g ; g \in T\}'$ is also trivial. But for this case we have also shown that

$U_g \in \pi(\mathbf{A(x)}) \; \forall g \in T$ and thus if E is a projection in $\pi(\mathbf{A(x)})'$ then clearly

$$E \in \pi(\mathbf{A(x)})' \cap \{U_g; g \in T\}' = \{\lambda I\}.$$

The GNS representation is thus irreducible in the sense of Murray-von Neumann, corresponding to f being a pure quantum state.

N = 1 Supersymmetry

For either case we need to add an additional element, normally denoted Q_α, to the Lie algebra, to represent the spread of spin values. In any representation, these are all linear operators, including the identity operator I, and thus form an algebra of such operators. Such a factorial representation corresponding to a set of particles, must contain equal numbers of bosons and fermions [9]. With certain assumptions, such a representation where the centre of the algebra is non-trivial can be decomposed into a direct integral of factorial representations, as discussed in [5].

We can develop a locally linear representation of these operators, which is a faithful representation of the Superspace Lie Algebra [9] by adding a pair of Grassmann variables to the algebraic formulation. We can then generate a standard Lie algebra while mixing commutators anticommutators. For example, if ξ and $\bar{\xi}$ have the Grassmann property so that $\xi\bar{\xi} = -\bar{\xi}\xi$,

then; $\left[\xi Q_A, \bar{\xi}\bar{Q}_{\dot{B}} \right] = \left(\xi Q_A \bar{\xi}\bar{Q}_{\dot{B}} - \bar{\xi}\bar{Q}_{\dot{B}}\xi Q_A \right) = \xi\bar{\xi}\left(Q_A\bar{Q}_{\dot{B}} + \bar{Q}_{\dot{B}}Q_A \right) = \xi\bar{\xi}\{Q_A,\bar{Q}_{\dot{B}}\}.$

Note that we also require that these Grassmann variables commute with the operators Q. A typical element of the corresponding Lie group G is then of the form;

$$G(x^\mu, \xi, \bar{\xi}) = \exp i(\xi Q + \bar{\xi}\bar{Q} - x^\mu P_\mu)$$

Here, x^μ is an event point in locally flat space-time, thus we can think of the Grassmann variables as a vector at the point x^μ. With this structure, the Superspace is a vector bundle and the locally flat group multiplication structure is of the following form;

$G(x^\mu, \theta, \bar{\theta})G(a^\nu, \xi, \bar{\xi}) = G(x^\mu + a^\nu - i\xi\sigma^\mu\bar{\theta} + i\theta\sigma^\mu\bar{\xi}, \theta + \xi, \bar{\theta} + \bar{\xi})$. This follows from the Grassmannian properties, since $\theta^2 = \bar{\theta}^2 = 0$, for example.

We can interpret this group product as shifting the locus of the Grassmann vector in space-time from x^μ to $x^\mu + a^\mu - i\xi\sigma^\mu\bar{\theta} + i\theta\sigma^\mu\bar{\xi}$ together with additive change to the Grassmann vector at this point. If this shift is infinitesimal, then, as in normal Lie group theory, we can consider the tangent plane around the group element $G(a^\mu, \xi, \bar{\xi})$, giving the following local representation of the Lie group generators on the tangent plane to the Riemannian space-time manifold at the event point (a^ν):

$$P_\mu(a^\nu) = i\frac{\partial}{\partial x^\mu}\Big|_{(a^\nu)} = i\partial_\mu\Big|_{(a^\nu)}; iQ_A(a^\nu) = \frac{\partial}{\partial\theta_A}\Big|_{(a^\nu)} -i\sigma^\mu\bar{\theta}\frac{\partial}{\partial x^\mu}\Big|_{(a^\nu)}; i\bar{Q}_{\dot{A}}(a^\nu) = -\frac{\partial}{\partial\bar{\theta}_{\dot{A}}}\Big|_{(a^\nu)} +i\theta\sigma^\mu\frac{\partial}{\partial x^\mu}\Big|_{(a^\nu)}$$

In this form the generators satisfy all the algebraic relationships of the Supersymmetric extension of the local translation Lie algebra. Thus, this representation is locally an algebraic isomorphism onto the curved Riemannian manifold M and the piecewise local representations;

$$P_\mu(a^\nu) = i\frac{\partial}{\partial x^\mu}\Big|_{(a^\nu)} = i\partial_\mu\Big|_{(a^\nu)}$$

are the generators of the local relativistic diffeomorphisms around the event points of M. The manifold is assumed smoothly differentiable; we focus on the flat tangent plane and assume Dirac relativistic spinor theory applies. We therefore assume that this operator representation acts on a 4-dimensional Hilbert space H of (spinor) wave functions; and we denote by ψ an element of the Hilbert space.

Within these assumptions we have the Dirac matrix $\gamma^5 = i\gamma^0\gamma^1\gamma^2\gamma^3$, and we can express this state ψ in the form $\psi = \frac{1}{2}(I_{4x4} - \gamma^5)\psi + \frac{1}{2}(I_{4x4} + \gamma^5)\psi$. We define these two terms as the left landed and right handed spinor components of the state vector, thus we define;

$$\psi_L = \frac{1}{2}(I_{4x4} - \gamma^5)\psi \text{ and } \psi_R = \frac{1}{2}(I_{4x4} + \gamma^5)\psi$$

So that $\psi = \psi_L + \psi_R$.

If we define the Dirac adjoint function, $\bar{\psi} = \psi^+\gamma^0$ where $\psi^+ = (\psi^*)^T$ then taking complex conjugates of both sides of the massless Dirac equation followed by transposition yields the identity:

$$(\psi^*)^T(-i(\gamma^{\mu*})^T(\partial_\mu + ieA_\mu) = 0$$

41

Exploiting the fact that $(\gamma^0)^2 = I_{4x4}$ leads to the equation:

$$-i\gamma^{\mu T}(\partial_\mu + ieA_\mu)\bar{\psi}^T = 0$$

The matrices $(-\gamma^{\mu T})$ also satisfy the Clifford algebra relations and there is a 4x4 non-singular matrix C such that $C^{-1}\gamma^\mu C = -\gamma^{\mu T}$. We can thus define the charge conjugate spinor

$$\psi^c = C\bar{\psi}^T .$$

In the Weyl representation we take;

$$C = i\gamma^0\gamma^2 = i\begin{pmatrix} -\sigma^2 & 0 \\ 0 & \sigma^2 \end{pmatrix} \text{ where } \sigma^2 \text{ is the second Pauli matrix } \begin{pmatrix} 0 & -i \\ i & 0 \end{pmatrix}$$

With ψ the spinor wave function $\begin{pmatrix} \varphi \\ \bar{\chi} \end{pmatrix} = \begin{pmatrix} \varphi_A \\ \bar{\chi}^{\dot{A}} \end{pmatrix}$ we have

$$\psi^c = C\bar{\psi}^T = C(\psi^+\gamma^0)^T = i\gamma^0\gamma^2\gamma^{0T}\psi^* = i\begin{pmatrix} 0 & -\sigma^2 \\ \sigma^2 & 0 \end{pmatrix}\begin{pmatrix} \varphi^* \\ \bar{\chi}^* \end{pmatrix} = i\begin{pmatrix} -\sigma^2\bar{\chi}^* \\ \sigma^2\varphi^* \end{pmatrix}$$

Now note that;

$\varphi^* = (\varphi_A)^* = \bar{\varphi}_{\dot{A}}$ and $\chi = \chi^A = (\bar{\chi}^{\dot{A}})^*$

Thus

$$\psi^c = i\begin{pmatrix} -\sigma^2\bar{\chi}^* \\ \sigma^2\varphi^* \end{pmatrix} = \begin{pmatrix} \begin{pmatrix} 0 & -1 \\ 1 & 0 \end{pmatrix}\bar{\chi}^* \\ \begin{pmatrix} 0 & 1 \\ -1 & 0 \end{pmatrix}\varphi^* \end{pmatrix} = \begin{pmatrix} \varepsilon_{AB}(\bar{\chi}^{\dot{A}})^* \\ \varepsilon^{\dot{A}\dot{B}}(\varphi_A)^* \end{pmatrix} = \begin{pmatrix} \varepsilon_{AB}\chi^A \\ \varepsilon^{\dot{A}\dot{B}}\bar{\varphi}_{\dot{A}} \end{pmatrix} = \begin{pmatrix} \chi_B \\ \bar{\varphi}^{\dot{B}} \end{pmatrix} \in H$$

where the ε matrices are the spinor metric 2x2 matrices.

The Mass Zero Case as a Graded Lie Algebra

We start with the properties of a Z_2-graded Lie algebra $L = L_0 \oplus L_1$ where L_0 is the Lie algebra spanned by the generators P_μ $(\mu = 0,1,2,3)$ and
L_1 is spanned by the spinor charge generators Q_α $(\alpha = 0,1,2,3)$.

From the definition of a Z_2-graded Lie algebra L;

$P_\mu \in L_0, Q_\alpha \in L_1$ with gradings 0 and 1

$(P_\mu, Q_\alpha) \rightarrow P_\mu \circ Q_\alpha = P_\mu Q_\alpha - (-1)^{0\times 1}Q_\alpha P_\mu = P_\mu Q_\alpha - Q_\alpha P_\mu = [P_\mu, Q_\alpha] = 0$

Similarly,

We also have $Q_\alpha \circ Q_\beta = Q_\alpha Q_\beta - (-1)^{1 \times 1} Q_\beta Q_\alpha = Q_\alpha Q_\beta + Q_\beta Q_\alpha = \{Q_\alpha, Q_\beta\}$, the anticommutator.

Since Q_α is a non-Hermitian operator (by construction), we can also consider the complex conjugate Dirac 4-spinor $\overline{Q}_\beta \in \dot{F}$. To resolve differences in the literature we assume $\overline{Q}_\beta = \dot{Q}_\beta$.

To constrain the number of options it is convenient at this stage to assume;

$Q = Q^c = C\overline{Q}^T$

Thus $(CQ)^T = (C^2 \overline{Q}^T)^T = -\overline{Q}$ since $C^2 = -1$.

Hence $Q^T C = \overline{Q}$ since $C^T = -C$

Following now the logic of [9] in general, we consider from the Z_2 grading,

$\{Q_\alpha, Q_\beta\} = Q_\alpha Q_\beta + Q_\beta Q_\alpha = a(\gamma^\mu C)_{\alpha\beta} P_\mu$

Multiplying from the right by C; $Q_\alpha Q_\beta C_{\beta\delta} + Q_\beta C_{\beta\delta} Q_\alpha = a(\gamma^\mu)_{\alpha\gamma} C_{\gamma\beta} C_{\beta\delta} P_\mu = -a(\gamma^\mu)_{\alpha\delta} P_\mu$

Hence $Q_\alpha (Q^T C)_\delta + (Q^T C)_\delta Q_\alpha = -a(\gamma^\mu)_{\alpha\delta} P_\mu$

Thus $\{Q_\alpha, \overline{Q}_\beta\} = -a(\gamma^\mu)_{\alpha\delta} P_\mu$

We assume the operators $Q, \overline{Q}$ are Marjorana 4-spinors; then in 2-spinor notation we can simply replace the Dirac γ matrices with their equivalent Pauli matrices yielding the following relationship;

$\{Q_A, \overline{Q}_{\dot{B}}\} - a(\sigma^\mu)_{A\dot{B}} P_\mu$ ....(2)

Similarly, we can show that, for 4-spinors, $\{\overline{Q}_\alpha, \overline{Q}_\beta\} = -a(C^{-1}\gamma^\mu)_{\alpha\beta} P_\mu$

Hence in 2-spinor notation, $\{\overline{Q}_A, \overline{Q}_{\dot{B}}\} = -a(C^{-1}\sigma^\mu)_{A\dot{B}} P_\mu$

Since all the latter equations are relativistically invariant, we can transform them to the rest frame where $P_\mu = (E,0,0,0) = (m,0,0,0)$ with the speed of light normalised at $c = 1$.

With these values of P_μ in equation (2) above we have

$\{Q_A, \overline{Q}_{\dot{B}}\} = -a\sigma^0_{A\dot{B}} P_0$

Hence $\{Q_A, \overline{Q}_{\dot{B}}\} \sigma^0_{\dot{B}A} = -a\sigma^0_{A\dot{B}} \sigma^0_{\dot{B}A} P_0 = -aP_0$

Taking $A = 1$, $\dot{B} = \dot{1}$ we have $\{Q_1, \overline{Q}_{\dot{1}}\} \sigma^0_{\dot{1}1} = -aP_0$

Similarly for $A = 2$, $\dot{B} = \dot{2}$ we have $\{Q_2, \overline{Q}_{\dot{2}}\} \sigma^0_{\dot{2}2} = -aP_0$

Since $\sigma^0_{\dot{1}1} = \sigma^0_{\dot{2}2} = 1$ we have $\{Q_1, \overline{Q}_{\dot{1}}\} + \{Q_2, \overline{Q}_{\dot{2}}\} = -2aP_0$

For consistency with the current literature, we assume the constant $a = -1$.

Thus, we have the quantum operator equality $\{Q_1, \bar{Q}_{\dot{1}}\} + \{Q_2, \bar{Q}_{\dot{2}}\} = 2P_0$

The left-hand side of this expression is a positive definite quantum operator thus for ψ an element of the Hilbert space;

If ψ is the vacuum state then $<\psi, P_0\psi> = 0$ is equivalent to $2 <\psi, Q_1\bar{Q}_{\dot{1}}\psi> + 2 <\psi, Q_2\bar{Q}_{\dot{2}}\psi> = 0$

Since, for 2-spinors, $\varphi_A^* = \bar{\varphi}_{\dot{A}}$ we can rewrite this as: $2 <\psi, Q_1 Q_1^*\psi> + 2 <\psi, Q_2 Q_2^*\psi> = 0$

Thus $2 \| Q_1\psi \|^2 + 2 \| Q_2\psi \|^2 = 0$, from which we deduce that $Q_1 |\psi> = Q_2 |\psi> = 0$

and also $Q_{\dot{1}} |\psi> = Q_{\dot{2}} |\psi> = 0 \Rightarrow \omega_\psi(Q_1) = \omega_\psi(Q_2) = \omega_\psi(Q_{\dot{1}}) = \omega_\psi(Q_{\dot{2}}) = = \omega_\psi(P_0) = 0$.

Factorial Representations of the Graded Lie Algebra

The extension of the space L_0 to the space $L = L_0 \oplus L_1$ maintains P^2 as an element of the centre;

$$\left[P^2, Q_A\right] = 0 = \left[P^2, \bar{Q}^{\dot{A}}\right]$$

In a factorial representation Π of the Z_2 graded algebra, it follows that $\Pi(P^2) = m^2 I$, fixing the mass m of all particles in this representation. However, the spins of the particles in this representation are not fixed at a common value.

In this factorial 2-spinor representation, we have the algebraic identity ;

$\{Q_A, \bar{Q}_{\dot{B}}\} = \sigma_{A\dot{B}}^\mu P_\mu$ from equation (A) with the parameter $a = -1$.

With $P^2 = m^2 I$ in this representation, and setting $P_\mu = (m, 0, 0, 0)^T$ in the rest frame, we have the following set of identities from equation (2) taking $A = 1, 2$ and $\dot{B} = \dot{1}, \dot{2}$;

$$\{Q_1, \bar{Q}_{\dot{1}}\} = \sigma_{1\dot{1}}^\mu P_\mu = \sigma_{1\dot{1}}^0 m = m$$
$$\{Q_1, \bar{Q}_{\dot{2}}\} = \sigma_{1\dot{2}}^\mu P_\mu = \sigma_{1\dot{2}}^0 m = 0$$
$$\{Q_A, Q_B\} = \{\bar{Q}_{\dot{A}}, \bar{Q}_{\dot{B}}\} = 0$$

From these properties we see that the 2-spinors form at least a Clifford algebra in this factorial representation, and we see also that $Q_A^2 = 0$ and this is in fact a Grassmann algebra.

If $| p, \lambda >$ is an eigenstate in the Hilbert space H, then $P_\mu | p, \lambda > = p_\mu | p, \lambda >$.

The corresponding (pure) vector state is invariant under the translational group since we have

$$0 = P_\mu Q_A | p, \lambda > = Q_A P_\mu | p, \lambda > = Q_A p_\mu | p, \lambda > = p_\mu Q_A | p, \lambda >$$

We can thus always choose a minimum energy pure vector state $\omega_{|p,\lambda>}$ which is translation invariant and with $\omega_{|p,\lambda>}(Q_A) = < p, \lambda | Q_A | p, \lambda > = 0$. It is specified by its mass-energy p and its spin value λ. Thus $\omega_{|p,\lambda>}$ is a translation invariant ergodic pure state.

If we now, exploiting local special relativistic covariance, choose an inertial reference frame in which the Wigner little group contains the spin generating 2x2 rotation matrices in the x-y plane, we have;

$$P_\mu = (E, 0, 0, E) \text{ therefore;}$$

$$\{Q_A, \bar{Q}_{\dot{B}}\} \mid p, \lambda >= \sigma^\mu_{A\dot{B}} P_\mu \mid p, \lambda >= \sigma^\mu_{A\dot{B}} P_\mu \mid p, \lambda >$$

$$\text{Now } \sigma^\mu_{A\dot{B}} P_\mu = \sigma^0 p_0 - \sigma^1 p_1 - \sigma^2 p_2 - \sigma^3 p_3 = E(\sigma^0 - \sigma^3)$$

$$= E\left(\begin{pmatrix} 1 & 0 \\ 0 & 1 \end{pmatrix} - \begin{pmatrix} 1 & 0 \\ 0 & -1 \end{pmatrix} \right) = \begin{pmatrix} 0 & 0 \\ 0 & 2E \end{pmatrix}$$

Hence applying this to our translation invariant vacuum state, $\omega_{|p,2>}$ we have;

$$|< p, 2 \mid Q_1 \bar{Q}_1 \mid p, 2 >= - < p, 2 \mid \bar{Q}_1 Q_1 \mid p, 2 >= 0$$

Similarly, considering the 2i element of the matrix, we have

$$|< p, 2 \mid Q_2 \bar{Q}_1 \mid p, 2 >= - < p, 2 \mid \bar{Q}_1 Q_2 \mid p, 2 >= 0$$

Thus $(aQ_1 + bQ_2)\bar{Q}_1 \mid p, 2 >= 0$ for any scalars a and b.

Since Q_1 and Q_2 span the subspace L_1, it follows that $\bar{Q}_1 \mid p, 2 >= 0$.

We conclude that the translation invariant pure Boson state $\omega_{|p,2>}$ is the local Clifford vacuum, and a spin-2 pure state corresponding to the graviton. It is an extreme point of the closed convex hull of the state space and is thus an extreme point of the translation invariant states: it is an ergodic state. Its Supersymmetric fermion partner is the Gravitino with spin $\frac{3}{2}$.

In Conclusion;

➤ From the properties of factorial representations of the translational group which are the generators of local diffeomorphisms and gravity; a minimal form of N = 1 Supersymmetry naturally emerges.

➤ Based on this theory, we predict a massless spin – 2 graviton and its supersymmetric spin - partner.

➤ The natural cutoff induced by the Supersymmetry assumptions implied that this theory is applicable in the low energy regime and may be within the parameters of observation of the LHC.

We define $\alpha : g \rightarrow \alpha_g$ to be a representation of T as automorphisms of the local fibre algebra

A(x) which we assume to be isomorphic to a von Neumann algebra with trivial centre acting on

a separable Hilbert space. The set $\{v(g) : g \rightarrow f \circ \alpha_g; f$ is a quantum state, and $g \in T\}$ is a

continuous group of commuting transformations of a subset of the dual space **A(x)***. If f is any

quantum state of the algebra, then define $\mathcal{E}$ to be the weak* closed convex hull of the set

$\{v(g)f; g \in T\}$. $\mathcal{E}$ is a weak* compact convex set and each $v(g) : \mathcal{E} \rightarrow \mathcal{E}$; by the Markov-

Kakutani fixed point theorem it follows that $\mathcal{E}$ has an invariant element.

The group representation α acts ergodically on **A(x)** if given a projection E in **A(x)**,

$\alpha_g(E) = E \ \forall g \in T$ implies that $E=0$ or $E=I$. Conversely, a quantum state f of A(x) is α-

invariant if $f(\alpha_g(A)) = f(A) \ \ \forall A$ in **A(x).** If f is an extreme point of the set of invariant states,

then f is defined to be a T-ergodic state. Note that the set of invariant states is a non-void weak*

compact convex subset of the generalised quantum state space of **A(x)** and is thus generated by

its extreme points. Since there is a non-trivial invariant state for the amenable group T, thus

there is an extremal T- invariant state of **A(x)**.

The fibre algebra **A(x)** is a quantum operator algebra and thus has an identity operator I. If f is a

quantum state of **A(x)** then by definition, $f(I)=1$. The support of f is the unique smallest

projection E in **A(x)** such that $f(E) =1$; i.e. $f(I-E) = 0$. Thus, if f is separating, then the support

of f is I.

We also require the following: E is an α-invariant projection in **A(x)** if $\alpha_g(E) = E \ \ \forall g \in T$.

It now follows that if f is separating, then it is an ergodic state if and only if the representation α

acts ergodically on **A(x)**. For let π be the Gelfand-Naimark-Segal (GNS) representation of

A(x) induced by the state f on the Hilbert space $H(f)$. Since f is separating $E_f = I$. In this case

π is a *-isomorphism, and $\pi \circ f = \omega_\xi$

Consider now the involution mapping on $\pi(\mathbf{A(x)})$ defined as $A \rightarrow A^*$. This induces an anti-

linear mapping on a dense subset of the Hilbert space $H(f)$; $S : A\xi \rightarrow A^*\xi$. Moreover, this

extends to a mapping with closed graph which we also denote by S. By the theorem of Tomita-

Takesaki [10] S has a polar decomposition $S = J\Delta^{\frac{1}{2}}$ such that $J\pi(A(x))J=\pi(A(x))$'; the commutant of the fibre algebra $\pi(A(x))$. If $x = B\xi$ is in the domain of S, then it follows that $U_g x$ also lies in the domain of S, and we have the relationship;

$$U_g SB\xi = U_g B^*\xi = \alpha_g(B^*)\xi = \alpha_g(B)^*\xi = S\alpha_g(B)\xi = SU_g B\xi$$

This leads to the conclusion that, on the domain of S, we have $U_g S = SU_g$.

Then we have;

$$S = U_g SU_g^* = U_g J\Delta^{\frac{1}{2}}U_g^* = U_g JU_g^* U_g \Delta^{\frac{1}{2}}U_g^*$$

By uniqueness of the polar decomposition, $J = U_g JU_g^*$; J and U_g commute for all $g \in T$. From this we deduce that;

$B \in \pi(A(x))\cap\{U_g; g \in T\}'$ implies that $JBJU_g = U_g JBJ$ for all $g \in T$.

Thus $JBJ \in \{U_g; g \in T\}' \cap \pi(A(x))'$.

Conversely, if $C \in \{U_g; g \in T\}' \cap \pi(A(x))'$ then $C = JBJ$ for some $B \in \pi(A(x))$ and $JBJU_g = U_g JBJ$ implies $JBU_g J = JU_g BJ$ and thus $BU_g = U_g B$ so that $B \in \{U_g; g \in T\}'$.

We conclude that;

$$J\left\{\pi(A(x))\cap\{U_g; g \in T\}'\right\}J = \pi(A(x))'\cap\{U_g; g \in T\}' \quad\text{.................(3)}$$

The automorphic representation $\alpha : g \rightarrow \alpha_g$ of T acts ergodically if and only if $\pi(A(x))\cap\{U_g; g \in T\}'$ is trivial, containing only the projections 0 and I and thus consisting of the set of complex multiples of I. From the reasoning above and noting that $J^2 = 1$, it follows that the representation $\alpha : g \rightarrow \alpha_g$ of T acts ergodically if and only if $\pi(A(x))'\cap\{U_g; g \in T\}'$ is also trivial.

We consider a compact Lie group G acting as a gauge group of automorphisms of the fibre algebra $\mathbf{A(x)}$. Recall that $\mathbf{A(x)}$ is a Quantum Operator Algebra; a von Neumann algebra with trivial centre acting (up to isomorphism) on a separable Hilbert space $F(x)$, and that locally the fibre bundle is a product bundle. We define a section algebra $O(D)$ as the closure in the ultraweak operator topology of the set of all such fibre algebras with the algebraic operations defined fibrewise $O(D) = \{\mathbf{A(x)}; \mathbf{x} \in D\}^{-\sigma w}$ with the base D a suitable subset of spacetime of physical interest. It can be characterised as the dual space of it predual $O(D)_*$ which consists of all normal, density matrix, states. Consider a representation of the compact Lie group G as automorphisms $\alpha : g \rightarrow \alpha_g(A)$ for $A \in O(D), g \in G$.

Given a partition of a locally curved space-time region into subsets $D(1), D(2), ..., D(n)$ we have from early foundational work [11] that;

$(O(D1) \times G) \otimes O(D2) \times G)$ is isomorphic to $(OD1 \otimes OD2) \times (G \times G = G^2)$. If we assume that by induction we have a similar result for the case n-1 then;

Applying this to the algebras $(\overset{n-1}{\underset{j=1}{\bigotimes}}(O(D(j)) \times G^{n-1}$ and $O(D(n)) \times G$ implies that;

$(\overset{n}{\underset{j=1}{\bigotimes}}(O(D(j)) \times G^n$ is isomorphic to $(O(D(1) \times G) \otimes ..... \otimes O(D(n) \times G)$

Hence by induction this expression must hold for all values of .n. As an example, let us we partition the event horizon D of a Black Hole as a subset of spacetime into n subsets $D(1), D(2), ..., D(n),$ then it follows, if we have a section algebra $O(D$ that in each subset we can locate a non-trivial section algebra $O(D(j))$. Choose a fibre algebra from each section algebra with a non-trivial state space which is generated by its extreme points – the pure states. This implies that corresponding to each partition element $D(j)$ we can associate a pure state $f(j)$.

In quantum ergodic theory, the partition of a space into measurable subsets corresponds to a level of information about location in the space and this is measured by the information entropy of the partition. This links the black hole event horizon surface area, partitions of that area, and measures of information entropy. We interpret the von Neumann entropy as an inverse measure of the amount of information that the quantum system in a given state will yield through measurement. The larger the entropy of the quantum system, the less information can be extracted.

This relates to the classical macro- level Clausius definition of entropy E; $\oint \frac{\delta Q}{T} = \Delta E$ where, integrated over a Carnot cycle, a change in energy δQ results in a change in entropy corresponding to a change in space-time curvature. [12].

The weak topology $\sigma\left(A(x)_*, A(x)\right)$ can be defined on the predual $A(x)_*$ as the coarsest topology for which elements of the predual are continuous. It is defined by a set of semi-norms $p = |f|$ for f a density matrix linear functional which as a set are separating for $A(x)_*$. Making minimal assumptions we let $\alpha : g \to \alpha_g$ be a weakly measurable representation of the compact Lie group G as automorphisms of $\mathbf{A(x)}$. By this we mean that the induced mapping[2] $v : g \to f \circ \alpha_g^{-1} : G \to \mathbf{A(x)}_*$ is measurable for Haar measure on G and the $\sigma\left(A(x)_*, A(x)\right)$ topology on $A(x)_*$ Since every positive element of $\mathbf{A(x)}_*$ is a countable sum of vector states this is equivalent to the definition that $v : g \to \omega_x \circ \alpha_g : G \to \mathbf{A(x)}_*$ is measurable for all x in the fibre Hilbert space $F(x)$. Given that the induced mapping $v : g \to f \circ \alpha_g : G \to \mathbf{A(x)}_*$ is measurable in the sense now defined above we have from [13];

$$\| v(g) \circ f(A) - v(h) \circ f(A) \| \leq \| v(g) \circ f - v(h) \circ f \| \| A \| \to 0 \text{ as } g \to h$$

This demonstrates the following result, which allows the extension of continuous gauge automorphic representations of compact Lie groups to their cross- product such as the Standard Model gauge group $SU(3) \times SU(2) \times S(1)$;

For the induced representation $v : g \to f \circ \alpha_g : G \to \mathbf{A(x)}_*$ *on the predual of A(x), weak measurability is equivalent to weak continuity.*

---

[2]We will ignore the inverse symbol in this definition to ease notational clutter.

49

We next show, as we did for local diffeomorphism-invariant quantum states [3, 8] that quantum states invariant under the action now of compact Lie groups are common in the sense that the weakly closed convex hull of every normal state contains such a state. We are now dealing with groups such as *SU(n)* which are both compact and non-abelian thus different techniques are required. To achieve this result, we have developed a new idea based on group stabilizer theory which we call Wigner sets. These are complementary to little groups.

Given a density matrix quantum state f, and a weakly measurable representation $g \to \alpha_g$ of a compact Lie group G as gauge automorphisms of the fibre algebra $\mathbf{A(x)}$; define the closed convex hull; $X(f) = \overline{co}\{f \circ \alpha_g; g \in G\} \subset \mathbf{A(x)}_*$ with closure in the $\sigma(A(x)_*, A(x))$ -topology. Define the group of isometric and $\sigma(A(x)_*, A(x))$ -continuous transformations mapping $X(f) \to X(f)$ by $v(G) = \{v(g) : x \to x \circ \alpha_g; g \in G, x \in X(f)\}$.

Mathematically, we note that since G is compact and $f \circ \alpha : G \to \mathbf{A(x)}_*$ is weakly measurable and thus weakly continuous; this implies that $f \circ \alpha(G)$ is $\sigma(A(x)_*, A(x))$ - compact. The Krein - Smulian theorem [14], then shows that

$X(f)$ is also a $\sigma(A(x)_*, A(x))$ - compact set. Thus $X(f)$ is a non-void $\sigma(A(x)_*, A(x))$ - compact convex subset of the locally convex Hausdorff linear topological space of ultraweakly continuous linear functionals acting on the fibre algebra $\mathbf{A(x)}$. The group of mappings $v(G) = \{v(g) : x \to x \circ \alpha_g; g \in G, x \in X(f)\}$ is, as we will now show, a non-contracting (semi)-group of weakly continuous affine maps of $X(f)$ onto itself.

If x, y are in $X(f)$ then;
$$\| v(g)x - v(g)y \| = \sup_{\|A\| \leq 1} \| v(g)x(A) - v(g)y(A) \| = \sup_{\|A\| \leq 1} \| x \circ \alpha_g(A) - y \circ \alpha_g(A) \| = \| x - y \|$$
since each α_g is a continuous bijection. Thus if $x \neq y$ and
$S = $ the weak closure of $\{v(g)x(A) - v(g)y(A); A \in \mathbf{A(x)}\}$ then $0 \notin S$.

Thus each mapping $v(g)$ is non-contracting. Also, each $v(g)$ is affine, since for A in $\mathbf{A(x)}$ and $0 \leq \lambda \leq 1$.

$$\{v(g) \circ (\lambda x + (1-\lambda)y)\}(A) = \lambda x(\alpha_g(A)) + (1-\lambda)y(\alpha_g(A)) = \lambda v(g) \circ x(A) + (1-\lambda)v(g) \circ y(A)$$

We can therefore, apply the Ryll-Nardzewski fixed point theorem [15] to establish the existence of an invariant normal state contained in $X(f)$.

The physical implications highlight the role of what we have termed Wigner sets.

Given $g \in G$, define the Wigner set of the mapping $v(g): X(f) \to X(f)$ as the stabiliser set;
$\mathcal{F}(v(g)) = \{x \in X(f); v(g) \circ x = x\}$.

More generally, given a finite subset $\{g(j) \in G; j =, 2,...n\}$ and corresponding mappings $\{v(g(j)); j =, 2,...n\}$, we can construct the affine mapping $\frac{1}{n}\left(\sum_j v(g(j))\right): X(f) \to X(f)$.

We then define $\mathcal{F}\{v(g(j)); j =, 2,...n\} \triangleq \mathcal{F}\left(\frac{1}{n}\left(\sum_j v(g(j))\right)\right)$

We assert that the following relationship between Wigner sets applies;

$$\bigcap_j \mathcal{F}(v(g(j))) = \mathcal{F}\left(\frac{1}{n}\left(\sum_j v(g(j))\right)\right) \dots\dots\dots\dots\dots\dots\dots\dots\dots(1)$$

Clearly if $x \in \bigcap_j \mathcal{F}(v(g(j))) \Rightarrow v(g(j)) \circ x = x \ \forall \ j = 1,..,n \Rightarrow \frac{1}{n}\left(\sum_j v(g(j))\right) \circ x = x$

$\Rightarrow x \in \mathcal{F}\left(\frac{1}{n}\left(\sum_j v(g(j))\right)\right)$ therefore $\bigcap_j \mathcal{F}(v(g(j))) \subseteq \mathcal{F}\left(\frac{1}{n}\left(\sum_j v(g(j))\right)\right)$

To complete the proof of equation (1) we now need to show that;

$$\mathcal{F}\left(\frac{1}{n}\left(\sum_j v(g(j))\right)\right) \subseteq \bigcap_j \mathcal{F}(v(g(j))) \dots\dots\dots\dots\dots\dots(2)$$

This relation is trivially true for n=1.

If the result (2) is false then;

There is a minimum positive integer $r \geq 2$ for which (2)
fails..............................(3)

For ease of exposition, denote the mapping $T(r) = \left(\frac{1}{r} \left(\sum_{j=1,..,r} v(g(j)) \right) \right)$ and the

mappings $T_j = v(g(j)), j = 1,.....,n$. so that $T(r) = \frac{1}{r} \left(\sum_{j=1,..,r} T_j \right)$ all acting on $X(f)$.

As noted earlier, since G is compact, the Krein-Smulian theorem, [14] shows that $X(f)$
is a $\sigma(A(x)_*, A(x))$ - compact set. $T(r)$ is an affine mapping on the compact convex set
$X(f)$ so by Schauder's extension of the Brouwer fixed point theorem [14] has a fixed
point $x(r)$ in $X(f)$. This implies that;

$$ T(r) \circ x(r) = \frac{1}{r} \left(\sum_{j=1,..,r} T_j \circ x(r) \right) = x(r). $$

By the definition of r as the minimum integer for which our assertion fails;

we must have an integer $k \leq r$ with $T_k \circ x(r) \neq x(r)$ ....(4)

Suppose now that $T_r \circ x(r) = x(r)$

Then

$$ rx(r) = rT(r) \circ x(r) = \left(\sum_{j=1,..,r} T_j \circ x(r) \right) = (T_1 + ... + T_{r-1}) \circ x(r) + T_r x(r) = (T_1 + ... + T_{r-1}) \circ x(r) + x(r) $$

$$ \Rightarrow x(r) = \frac{1}{(r-1)} (T_1 + ... + T_{r-1}) \circ x(r) \Rightarrow x(r) \in \mathcal{F}(T(r-1)) = \bigcap_{j=1,...,r-1} \mathcal{F}(v(g(j))) $$

$$ \Rightarrow T_j x(r) = x(r) \text{ for all } j = 1...,r $$

But by definition of k at (4) we have a logical contradiction.

By reordering the sum $\left(\sum_{j=1,..,r} T_j \circ x(r) \right)$ we can relabel r by any $j=1,...,r.$.

In summary, we have shown that if our assertion at equation (2) is false then;

52

$$T_j \circ x(r) \neq x(r) \text{ for all } j = (1, ..., r) \dots\dots\dots(5)$$

Let $\mathcal{S}$ be the subgroup generated by the finite set $\left\{T_j = \nu(g_j); j = (1, ..., r)\right\}$ and let $K(r)$ be the weakly closed convex hull of $\left\{S\, x(r); S \in \mathcal{S}\right\}$. Since the set of mappings $\nu(G)$ is non-contracting, and $T_j x(r) \neq x(r)$ for all $j = (1, ..., r)$ there is a continuous semi-norm p on $X(f)$ with;

$$p\left(ST_j x(r) - Sx(r)\right) > \varepsilon \ \forall\, j = (1, ..., r), S \in \mathcal{S} \dots\dots\dots(6)$$

We will prove that this implication of (3) also gives rise to a contradiction, exploiting now part of the argument in [16].

Since $\mathcal{S}$ is finitely generated, it follows that $K(r)$ is separable, as well as being a weakly compact, convex subset of the dual space of **A(x)**. It has the appropriate geometric and topological properties. Thus, for the given $\varepsilon > 0$ above there is a proper closed convex subset C of $K(r)$ with ;

$$\text{the } p\text{-diameter of } K(r) \backslash C = \sup\left(p(x - y); x, y \in K(r) \backslash C\right) \leq \varepsilon.$$

$K(r) \backslash C$ is non-void, and $K(r)$ is the weakly closed convex hull of $\left\{S\, x(r); S \in \mathcal{S}\right\}$, thus we can choose a mapping S such that $Sx(r) \in K(r) \backslash C$. Now our chosen S is an affine mapping, therefore;

$$Sx(r) = ST(r)x(r) = \frac{1}{r}\left(\sum_j ST_j x(r)\right)$$

C is a convex set, thus not every $ST_j x(r)$ can be a member of C, otherwise the expression above would then imply that $Sx(r) \in C$. Thus, for some j,

$ST_j x(r) \in K(r) \backslash C$.

We thus have;

$Sx(r) \in K(r) \backslash C$ and $ST_j x(r) \in K(r) \backslash C$. But for the seminorm p, $p - diam\left(K(r) \backslash C\right) \leq \varepsilon$.

Thus, for some j ;

$$p\left(ST_j x(r) - Sx(r)\right) \le \varepsilon.$$

This contradicts (6), showing that there is no minimum positive integer $r \ge 2$ for which (2) fails. and we have proved our assertion on the relation between Wigner sets;

$$\bigcap_j \mathcal{F}\left(\nu\left(g(j)\right)\right) = \mathcal{F}\left(\frac{1}{n}\left(\sum_j \nu\left(g(j)\right)\right)\right)$$

Invariant Normal States

It is now easy to prove that $X(f)$ contains a fixed point for the group of isometric and $\sigma\left(A(x)_*, A(x)\right)$-continuous transformations $v(G) = \left\{v(g)(x) = x \circ \alpha_g; g \in G, x \in X(f)\right\}$.

We have that $\left(\frac{1}{n}\left(\sum_j v\left(g(j)\right)\right)\right)$ is a $\sigma\left(A(x)_*, A(x)\right)$- continuous affine mapping on the compact convex set $X(f)$ it has a fixed point x (applying again Schauder's fixed point theorem). Then $x \in \mathcal{F}\left(\frac{1}{n}\left(\sum_j v\left(g(j)\right)\right)\right)$.

The expression;

$$\bigcap_j \mathcal{F}\left(\nu\left(g(j)\right)\right) = \mathcal{F}\left(\frac{1}{n}\left(\sum_j \nu\left(g(j)\right)\right)\right)$$

Shows that the Wigner sets $\mathcal{F}\left(v(g)\right)$ have the finite intersection property since $\bigcap_j \mathcal{F}\left(v\left(g(j)\right)\right)$ is non-void . Clearly each Wigner set is a $\left(\sigma\left(A(x)_*, A(x)\right)\right)$ closed subset of the compact set $X(f)$ thus $\bigcap_g \mathcal{F}\left(v(g)\right) \ne \emptyset$. If

$h \in \bigcap_g \mathcal{F}\left(v(g)\right) \Rightarrow h = v_g \circ h = h \circ \alpha_g \forall g \in G$. Thus, h is the required invariant quantum state.

The proof shows that quantum states invariant under the action of compact Lie groups are common in the sense that the weakly closed convex hull of ever normal state contains such an invariant state.

Chapter 5: Quantum Ergodic Theory and Black Hole Dynamics

Set Partitions as Location Information

Let p be a partition of Z into measurable subsets $\{A_j; j = 1, 2, 3.....\}$.

If Z is a compact measure space with total measure equal to 1 then we can interpret the value $\mu(A_j)$ as the probability of the set A_j. The expression $-\log \mu(A_j)$ is then a measure of the description length or Kolmogorov Complexity of the partition subset A_j. A measure preserving transformation such as T transforms the partition p into the partition Tp which is $\{TA_j; j = 1, 2, 3....\}$.

The *entropy or, equivalently, the expected value of the Kolmogorov Complexity of the partition p is defined as* $-\sum_j \mu(A_j)\log\mu(A_j)$.

Since T is measure preserving, the partitions p and Tp clearly have the same entropy.

Hopf [20] considers a partition of a measurable set into subsets and the effect of multiples T^k of T acting on those subsets. In operator theoretic language we can express this as follows. We replace measurable sets by projections and the group $\{T^k; k \in Z\}$ by a general discrete group G of automorphisms of a *commutative* von Neumann algebra R represented via the normal representation $\pi = \sum_{f \in S} \pi_f$, being the sum of the GNS representations for the set of all normal states S, acting on a Hilbert space H. The representation π is an isomorphism and is G – invariant. As shown earlier it follows that G can be implemented by a non-topological discrete unitary representation $g \to U_g$ from G to the set of unitaries acting on H.

We define two projections H and K in R to be *Hopf equivalent* if there is an orthogonal family of projections E_j in R and group elements $g_j \in G$ with $H = \sum_j E_j$ and $K = \sum_j U_{g_j}^* E_j U_{g_j}$. This equivalence leads to a partition based criterion, 'H-finiteness', for the existence of a finite invariant measure. For this kind of orthogonal partition we can define the entropy as being derived from the partition weightings (all equal in this case). With this definition it is then clear that if the two projections H and K in R are Hopf equivalent then they have the same entropy relative to an invariant measure. This idea is easily extended to the noncommutative case,

The Structure of Local Quantum Algebras

For each event point **x** in curved space-time, we assume we have a fibre algebra **A(x)** defined as a von Neumann algebra with trivial centre and a faithful representation as an algebra of operators acting on a separable Hilbert space.

Given a compact subset D of the analytic manifold of space-time M, such as the closure of a time-bounded, oriented double-cone, the local quantum algebra $O(D)$ is an associative fibre sub-bundle. It is *associative* in the sense that a Lie group (the translation subgroup of the Poincare group) acts as a local gauge group of automorphisms of each fibre algebra **A(x)**. It is a sub-bundle in the sense that $O(D) = \{A(x);\ x \in D\} \subset \{A(x);\ x \in M\}$.

The local von Neumann algebra $O(D)$ does not necessarily have a trivial centre; its structure is more complex than that of a fibre algebra. We assume that the quantum system it represents has an energy operator with discrete countable eigenstates and we thus assume also that $O(D)$ is separable. We propose to develop the ideas of noncommutative information theory to gain insight into the structure of $O(D)$, as we now describe.

Von Neumann introduced the idea of equivalence of measurement 'projection' operators as a way of gaining traction on the structure of a general von Neumann algebra, see for example [10]. Much of this analysis centers around the question of whether the algebra possesses a finite trace, extending the idea of the trace of a finite matrix operator as the sum of its observable eigenvalues. This analysis was enhanced to take account of groups of unitarily implemented automorphisms of the algebra by Stormer [19]. This allows him to define a 'G-equivalence' of projections which generalizes to the non-commutative quantum case the definition Hopf [20] used in standard commutative ergodic theory. One of us, Moffat [11], extended this work to a characterisation of the tensor product of 'G-type III' algebras.

As a result of this previous work we can now develop a classification of the structure of our local algebra $O(D)$. We do this by applying these earlier results where the group concerned is now the subgroup T of local translations of space-time of the Poincare group P, corresponding to local diffeomorphisms of the analytic sub-manifold D.

Let $\alpha : g \to \alpha_g$ be a group representation of the translation subgroup T of the Poincare group as a discrete group acting on the von Neumann algebra $O(D)$. A representation π of $O(D)$ acting on a Hilbert space H is covariant if there is a homomorphism $g \to U_g$ from T to the group of unitary operators on H with $\pi(\alpha_g(A)) = U_g \pi(A) U_g^* \quad \forall A \in O(D)$.

If φ is a normal state of $O(D)$ then $\varphi \circ \alpha_g$ is also a normal state since each automorphism preserves the algebraic structure and hence preserves complete additivity. If S denotes the set of all normal states of $O(D)$ then the direct sum $\pi = \oplus\{\pi_\varphi ; \varphi \in S\}$ of their Gelfand-Naimark-Segal (GNS) representations is a faithful representation of $O(D)$ as a von Neumann algebra acting on a Hilbert space H which is the direct sum of the GNS Hilbert spaces. If we define

$$U_g \left(\oplus_{\varphi \in S} \pi_\varphi(A_\varphi) x_\varphi \right) = \oplus_{\varphi \in S} \pi_\varphi(\alpha_g(A_{\varphi \circ \alpha_g}) x_\varphi$$ as a mapping on each of the pre-Hilbert spaces for the

GNS constructions, then U_g extends to a unitary operator on H and the representation

$\pi = \oplus\{\pi_\varphi ; \varphi \in S\}$ is a faithful normal representation of $O(D)$. We can therefore assume that T acting on $O(D)$ as a discrete group of automorphisms is unitarily implemented.

More formally, let $\alpha : g \to \alpha_g$ be a group representation of the translation subgroup T of the Poincare group as a discrete group acting on the von Neumann algebra $O(D)$. If E and F are projections in $O(D)$ we say that E and F are T-equivalent if there is a set of operators $\{A_g ; g \in T, A_g \in O(D)\}$ with $E = \sum_g A_g^* A_g$ and $F = \sum_g \alpha_g\left(A_g A_g^*\right)$. We write this T-equivalence

as $E \approx F$ and call it a T-twisted equivalence. In the special case that each A_g is a projection, this definition is a direct non-commutative generalisation of Hopf equivalence.

We define a projection F to be T-finite if F contains no proper sub-projections which are T-equivalent to F. The algebra $O(D)$ is defined to be T-finite, or T-Type II(1), if the identity of $O(D)$ is a T-finite projection. $O(D)$ is T-semifinite, or T-Type II(∞) if every projection in O(D) dominates a T-finite projection. $O(D)$ is T-purely infinite, or T-Type III, if $O(D)$ does not contain any T-finite projections.

The T-type III case is the most difficult to analyse. In the T-type III case there is not even the 'shadow' of a trace; a T-invariant trace being a bounded faithful normal linear mapping τ: $O(D) \to \mathbb{C}$ with;

$$\tau(AB) = \tau(\alpha_g(AB)) = \tau(BA) \quad \forall g \in T; A, B \in O(D).$$

If τ is a trace, then by our earlier remarks we can assume that the group representation of T, as a discrete group, is unitarily implemented by the unitary representation $U : g \to U_g$ and τ is automatically T-invariant. Stormer [19] established that $O(D)$ is T-semifinite if and only if there is a faithful normal semifinite T-invariant trace on $O(D)$.

In the commutative case, a general von Neumann algebra R is isomorphic to the set $L^\infty(Z, v)$ of essentially bounded, measurable, complex-valued functions on the locally compact set Z, with v a positive regular Borel measure. If G is a group of automorphisms of R then G is isomorphic to a group of automorphisms of $L^\infty(Z, v)$ which we also denote as G.

Projections P in the algebra R become, under this isomorphism, characteristic functions of Borel subsets of Z. If $g: P \to Q$ and P is isomorphic to χ_E for the Borel set E then Q is also a projection thus is isomorphic to χ_F for some Borel subset F. Thus the group G induces a group of transformations of the σ-ring $\mathcal{B}$ of Borel subsets of Z.

Let T be such a transformation which is measure-preserving, then for any Borel set E in $\mathcal{B}$, $v(T^{-1}(E)) = v(E)$ i.e. the measure v is T-invariant.

T is defined to be *ergodic* if $T^{-1}(E) = E$ modulo a null set implies either $v(E) = 0$ or $v(Z \setminus E) = 0$. If T is ergodic in this sense and the measure v is T-invariant, then v is defined to be an ergodic measure [21].

It then follows that if Z is compact, v is a probability measure on Z which is ergodic, and T and its inverse are continuous mappings, then this is equivalent to v being an extreme point of the invariant measures on Z.

To prove this, let $\mathcal{E}$ be the set of all invariant probability measures on Z, and let v be an extreme point of $\mathcal{E}$, then any measurable set E of positive measure with $0 < \lambda = v(E) < 1$ allows the construction of measures;

$$\left\{ \mu_1(K) = \frac{1}{\lambda} v(K \cap E) \text{ and } \mu_2(K) = \frac{1}{1-\lambda} v\big(K \cap (Z \setminus E)\big); K \subset Z \right\}$$

such that $v = \lambda \mu_1 + (1-\lambda)\mu_2$; a contradiction.

Conversely, if v is a probability measure on Z which is ergodic, and $0 < \mu < v$, then we have, for any Borel set A, $\mu(A) = \int_A f(x)dv(x)$ for some $f \in L^1(Z,v)$ and f is a T- invariant, positive function by the properties of probability measures. If f is not constant we can define Borel sets $S_1 = \{x \in Z; f(x) < t\}$ and $S_2 = \{x \in Z; f(x) > t\}$ for some positive real t. Both sets are invariant and non-trivial, thus they must both have measure 1; a contradiction. Hence f is constant and the measure is an extreme point since for any convex combination of measures, v dominates these measures and this leads to the tautology;

$$v = \lambda v + (1-\lambda)v \text{ for some } \lambda: 0 < \lambda < 1.$$

proving the result.

Non-commutative Information Theory

If $A(x)$ is a fibre algebra, let f be a faithful normal state of $A(x)$ and define the 'induced' mapping; $v_g(f) = f \circ \alpha_g$ This means that $v_g(f)$ is also a normal state and $v_g(f)(A) = f(\alpha_g(A))$.

Since the subgroup T is abelian, and the mapping $g \to v(g)$ is a group homomorphism, the set $\{v(g); g \in T\}$ is a continuous group of commuting transformations of the dual space $A(x)^*$. If f is a state of the algebra then define $\mathscr{E}$ to be the weak* closed convex hull of the set $\{v(g)f; g \in T\}$. $\mathscr{E}$ is a weak* compact convex set and each $v(g): \mathscr{E} \to \mathscr{E}$; by the Markov-Kakutani (MK) fixed point theorem [22], it follows that $\mathscr{E}$ has an invariant element. In other words, the group T has the fixed point property [23] and thus is amenable. In summary, because T is an abelian group and locally compact it is an amenable group, since the closed convex hull of any quantum state of the system contains a T-invariant state. This leads us to define the following;

Let $A(x)$ be a fibre algebra and the group T of translations of space-time a subgroup of the Poincare group. Let $\alpha: g \to \alpha_g$ be a representation of T as automorphisms of $A(x)$. The group representation α acts ergodically on **A(x)** if given a projection E in **A(x)**, $\alpha_g(E) = E \ \forall g \in T$ implies that $E=0$ or $E=I$.

This definition is a direct generalisation of the commutative case where **A(x)** is the set of essentially bounded measurable functions on a locally compact space with a regular Borel measure, discussed above. We can also have the following non- commutative generalisation of an ergodic probability measure as an extreme point of the set of invariant measures; as first pointed out by Segal [24].

Let **A(x)** be a fibre algebra and $\alpha : g \to \alpha_g$ a representation of the translation subgroup T of the Poincare group P as automorphisms of **A(x)**. A quantum state f of A(x) is α-invariant if $f(\alpha_g(A)) = f(A) \quad \forall A$ in **A(x)**. If f is a normal (i.e. density matrix) state and an extreme point of the set of invariant states, then f is defined to be a T-ergodic state.

Note that the set of invariant states is a compact convex subset of the generalised quantum state space of **A(x)** and is thus generated by its extreme points [25]. There is a non-trivial invariant state for the amenable group T, as discussed earlier, thus there is an extremal T- invariant state of **A(x)**. Since, by definition, the Hilbert space representation on which **A(x)** acts is separable, the algebra contains a faithful normal state and hence a T-invariant normal state. The norm limit of a set of normal states is again normal [10] and thus;

The fibre algebra A(x) with the assumptions above always contains a T-ergodic state

The Crossed Product Algebra of O(D)

Assume (by taking a faithful representation if necessary) that *O(D)* acts on a Hilbert space H. Define the Dirac function ε_g to take the value 1 at g and zero elsewhere on T. Then $\{\varepsilon_g ; g \in T\}$ is an orthonormal basis for the Hilbert space $l^2(T)$. Given $l^2(T)$ and H we can form the tensor product Hilbert space $H \otimes l^2(T)$. Define;

$$U_h(x \otimes \varepsilon_g) = x \otimes \varepsilon_{gh^{-1}} \text{ for } x \in H, \ g,h \in T; \Phi(A)(x \otimes \varepsilon_g) = \alpha_g(A)x \otimes \varepsilon_g \text{ for } A \in O(D), g \in T$$

Then U_h extends to a unitary operator on $H \otimes l^2(T)$ and the mapping U is a group homomorphism from the translation group T into the group of unitaries acting on $H \otimes l^2(T)$.

Similarly $\Phi(A)$ extends to a bounded linear operator on $H \otimes l^2(T)$ for all A in $O(D)$ and the mapping $U: h \to U_h$ implements the automorphic representation $h \to \alpha_h$.

The transformation Φ is an ultraweakly continuous *isomorphism of $O(D)$ and it follows that $\Phi(O(D))$ is a von Neumann algebra. Finite sums $\sum_j U_{g_j} \Phi(A_j)$ form a *algebra which contains $\Phi(O(D))$. The cross product algebra $O(D) \times T$ is defined as the closure of this *algebra for the ultraweak operator topology. The crossed product algebra can be used to prove the following structural result.

Assume $O(D1)$ and $O(D2)$ are local von Neumann algebras in space-time regions $D1$ and $D2$ which are not space-like separated. Let G and H be discrete representations of the translation subgroup of the Poincare group as automorphisms of $O(D1)$ and $O(D2)$ respectively. Then if either $O(D1)$ or $O(D2)$ is G/H-purely infinite (G/H-Type III), the tensor product algebra $O(D1) \otimes O(D2)$ is $G \times H$- purely infinite (equivalently $G \times H$-type III) under the action of the joint representation G×H of the translation group. If both O(D1) and O(D2) are G/H finite or G/H semifinite, then the same applies to the tensor product algebra. These results follow from the fact that [11]; $(O(D1) \times G) \otimes O(D2) \times H)$ is isomorphic to $(OD1 \otimes OD2) \times (G \times H)$..

Structural Symmetry

In this part of our analysis of the structure of $O(D)$ we derive a symmetry for purely infinite type III algebras between the T-type of $O(D)$ and the corresponding Murray-von Neumann type of its cross product algebra $O(D) \times T$. First we have to prove a key result.

Recall that, by construction, the crossed product von Neumann algebra O(D) ×T contains the embedded closed sub-algebra $\Phi(O(D))$, isomorphic to $O(D)$. Then it follows that there is an ultraweakly continuous mapping, denoted Γ, from O(D) ×T to O(D) such that the composite map $\Gamma \circ \Phi : O(D) \times T \to \Phi(O(D))$ is a continuous projection of norm one.

Proof. Continuing with the notation introduced earlier; the map $x \to x \otimes \varepsilon_g : H \to H_g$ is both isometric and linear. The Hilbert space $K = H \otimes l^2(T)$ is the direct sum of the H_g and every element x of K can be represented as. $x = \sum_{g \in T} x_g \otimes \varepsilon_g$ with $\| x \|^2 = \sum_{g \in T} \| x_g \|^2 < \infty$

If E_g is the projection from K onto H_g, and $B = \sum_g U_g \Phi(A_g)$ is an element of $(O(D) \times T)_0$ then

straightforward arguments show that $E_s B E_t = E_s U_{s^{-1}t} \Phi(A_{s^{-1}t}) E_t$.

Taking the weak closure, we have $B \in O(D) \times T$ with $B = \lim_\alpha \left\{ B^\alpha ; E_s B^\alpha E_t = E_s U_{s^{-1}t} \Phi(D^\alpha_{s^{-1}t}) E_t \right\}$

. From the Kaplansky density theorem [10] we can choose $\left\{ B^\alpha ; \| B^\alpha \| \leq \| B \| \right\}$ and the net $D^\alpha_{s^{-1}t}$

is then a bounded net in the ultraweakly compact ball of radius $\| B \|$. It thus has a subnet

converging to an element $D_{s^{-1}t}$ of $O(D)$. From this we have the following expression;

$$E_s B E_t = E_s U_{s^{-1}t} \Phi(D_{s^{-1}t}) E_t$$

In particular we have $E_e B E_e = E_e \Phi(D_e) E_e$. We define $\Gamma(B) = D_e$

Clearly the mapping Γ is linear, ultraweakly continuous, and $\Gamma|_{\Phi(O(D))} = \Phi^{-1}$.

If B is in the kernel of Γ then $D_e = 0$. From above this implies that $E_s B E_s = 0 \quad \forall s$ and thus B

$=0$; the kernel of Γ is $\{0\}$ and Γ is a faithful mapping. This shows that Γ has the required

properties.

This allows us to now prove the following key structural result.

O(D) is *T*-type III if and only if the crossed product algebra *O(D)* ×*T* is type III in the sense of

Murray-von Neumann.

Proof. If *O(D)* is not *T*-type III then it contains a non-trivial *T*-finite projection E. Then it

follows that if Φ is the identification of *O(D)* within the crossed product algebra *O(D)* ×*T* then

Φ(E) is finite in the sense of Murray-von Neumann. Thus *O(D)* ×*T* is not type III.

Conversely assume the crossed product algebra *O(D)* ×*T* is not type III. We proved that there is

a faithful normal projection Γ of norm one from *O(D)* ×*T* onto *O(D)*. It is then easy to see that

O(D) cannot be type III see reference [10]. Thus *O(D)Z* is semifinite for some projection Z in

the centre of *O(D)*: *O(D)* cannot be *T*-type III. This completes the proof.

Quantum Gravity States and Wandering Projections

The fibre algebra **A(x)** is a quantum operator algebra and thus has an identity operator *I*. If *f* is a

quantum state of **A(x)** then by definition, *f(I)=1*. The support of *f* is the unique smallest projection E in $\mathbf{A(x)}$ such that $f(E) = 1$ and is denoted E_f. Now let $\mathbf{A(x)}$ be a fibre algebra and $\alpha : g \rightarrow \alpha_g$ a representation of the translation subgroup T of the Poincare group P as gauge automorphisms of $\mathbf{A(x)}$. Assume there exists at least one state f of $\mathbf{A(x)}$ which is α-invariant. Then the support of f, E_f, is an invariant projection and f is a normal and ergodic state if and only if the representation α acts ergodically on the cut down algebra $E_f \mathbf{A(x)} E_f$.

Proof. We start with the observation that assuming f is α-invariant implies that $f(\alpha_g(E)) = f(E) \quad \forall E;g$; we call such states gravity states. By uniqueness of the support of f it follows that E_f is an α-invariant projection. Let π be the Gelfand-Naimark-Segal (GNS) representation of $\mathbf{A(x)}$ induced by the state f on the Hilbert space $H(f)$. We make the simplifying assumption for now that f is a faithful state; i.e. $E_f = I$, and revisit this assumption later. In this case the von Neumann algebra $\pi(\mathbf{A(x)})$ has a separating-generating vector ξ and the representation π is a *-isomorphism. Define the unitary group $U_g \pi(A) \xi = \pi\big(\alpha_g(A)\big)\xi$ on a dense subset of $H(f)$, then U_g extends to a unitary on $H(f) = \{\pi(B)\xi; B \in \mathbf{A(x)}\}^-$ with closure of the set in the norm topology. The mapping $U : g \rightarrow U_g$ is then a unitary representation of the translation group T and for B a quantum observable in the fibre algebra $\mathbf{A(x)}$ we have $U_g \pi(B) U_g^* = \pi(\alpha_g(B)) \quad \forall g \in T$ i.e. the unitary representation U implements the automorphic representation $\alpha : g \rightarrow \alpha_g$.

Consider now the involution mapping on $\pi(\mathbf{A(x)})$ defined as $A \rightarrow A^*$. This induces an anti-linear mapping on a dense subset of the Hilbert space $H(f)$; $S : A\xi \rightarrow A^*\xi$. Moreover, this extends to a mapping with closed graph which we also denote by S. By the theorem of Tomita-Takesaki [10] S has a polar decomposition $S = J\Delta^{\frac{1}{2}}$ such that $J\pi(\mathbf{A(x)})J = \pi(\mathbf{A(x)})'$; the commutant of the fibre algebra $\pi(A(x))$. If $x = B\xi$ is in the domain of S, then it follows that $U_g x$ also lies in the domain of S, and we have the relationship;

$$U_g S B\xi = U_g B^*\xi = \alpha_g(B^*)\xi = \alpha_g(B)^*\xi = S\alpha_g(B)\xi = SU_g B\xi$$

This leads to the conclusion that, on the domain of S, we have $U_g S = SU_g$.

Then we have;

$$S = U_g S U_g^* = U_g J \Delta^{\frac{1}{2}} U_g^* = U_g J U_g^* U_g \Delta^{\frac{1}{2}} U_g^*$$

By uniqueness of the polar decomposition, $J = U_g J U_g^*$; J and U_g commute for all $g \in T$. From this we deduce that;

$B \in \pi(A(x)) \cap \{U_g ; g \in T\}'$ implies that $JBJU_g = U_g JBJ$ for all $g \in T$.

Thus $JBJ \in \{U_g ; g \in T\}' \cap \pi(A(x))'$.

Conversely, if $C \in \{U_g ; g \in T\}' \cap \pi(A(x))'$ then $C = JBJ$ for some $B \in \pi(A(x))$ and

$JBJU_g = U_g JBJ$ implies $JBU_g J = JU_g BJ$ and thus $BU_g = U_g B$ so that $B \in \{U_g ; g \in T\}'$.

We conclude that $J \left\{ \pi(A(x)) \cap \{U_g ; g \in T\}' \right\} J = \pi(A(x))' \cap \{U_g ; g \in T\}'$.

The automorphic representation $\alpha : g \to \alpha_g$ of T acts ergodically if and only if

$\pi(A(x)) \cap \{U_g ; g \in T\}'$ is trivial, containing only the projections 0 and I and thus consisting of the set of complex multiples of I. From the reasoning above, and noting that $J^2 = 1$, it follows that the representation α of T acts ergodically if and only if $\pi(A(x))' \cap \{U_g ; g \in T\}'$ is also trivial.

If E is a projection in the set;

$$\pi(A(x))' \cap \{U_g ; g \in T\}'$$

we can define a state;

$$f_E(A) = \frac{\langle \xi, E\pi(A)\xi \rangle}{\langle \xi, E\xi \rangle}$$

on the fibre algebra $\mathbf{A}(x)$. Then $f_E = \omega_{E\xi} \circ \pi$ is a state dominated by $f = \omega_\xi \circ \pi$ and we have;

$$f(A) = \frac{\langle \xi, \pi(A)\xi \rangle}{\langle \xi, \xi \rangle} = \lambda \frac{\langle \xi, E\pi(A)\xi \rangle}{\langle \xi, E\xi \rangle} + (1-\lambda) \frac{\langle \xi, (I-E)\pi(A)\xi \rangle}{\langle \xi, (I-E)\xi \rangle} \quad \text{for } A \in \mathbf{A}(x),$$

$$\text{where } \lambda = \frac{\langle \xi, E\xi \rangle}{\langle \xi, \xi \rangle} = \frac{\| E\xi \|^2}{\| \xi \|^2} \text{ and } 1-\lambda = \frac{\| \xi \|^2 - \| E\xi \|^2}{\| \xi \|^2} = \frac{\langle \xi, (I-E)\xi \rangle}{\| \xi \|^2}$$

Thus f is an extremal invariant state if and only if the projection $E=0$ or I. The result follows for the support of f equal to 1. Finally, we need to extend the result to a general invariant state f with support E_f, $0 < E_f < I$. This follows from what we have already proved, since the restriction of f to $E_f \mathbf{A}(x) E_f$ is a faithful state, and a state extremal among the invariant states of the cut down algebra $E_f \mathbf{A}(x) E_f$ is also extremal among the invariant states of the full fibre algebra $\mathbf{A}(x)$ – hence no spontaneous symmetry breaking. This follows from the fact that if f is a convex combination of states from the full fibre algebra, then each of them has a support less than or equal to E_f.

In the next section we develop and prove a noncommutative version of a well-known result in classical ergodic theory and use it to characterise the existence of such symmetric 'gravity' states.

Hajian and Kakutani [6] defined a wandering set as follows. Let (X, B, μ) be a measure space with finite measure; $\mu(X) < \infty$ and where B is the set of all measurable subsets of X. Let T be a bijective transformation of X such that both T and its inverse are measurable mappings. A wandering set for T is a measurable subset S of X such that the sets $\{T^{nk}(S)\}$ are disjoint, for some infinite sequence of integers nk.

Recall that measures ν and μ on the measure space X are said to be equivalent if they share the same null sets. A measure ν is T-invariant if $\nu(T(E)) = \nu(E)$ for all measurable subsets E of X.

With these definitions, Hajian and Kakutani [6] showed that there is a finite T-invariant measure ν on X, equivalent to μ, if and only if there are no wandering subsets of X.

If now we consider an abelian von Neumann algebra R, then R is isomorphic to $C(X)$ with X a compact Stonean space of finite measure, and the positive, normal, regular Borel measures on X correspond to the normal states of R. By [17] we can characterise these *normal measures* as being equivalent to measures which annihilate each nowhere dense subset of X. It follows that if measures v and μ on the measure space X are equivalent and measure v is normal, then measure μ is also normal.

If θ is a continuous automorphism of the abelian algebra R, isomorphic to $C(X)$, then we can define the homeomorphism T of X by $\theta f(x) = f(Tx)$ for $f \in C(X)$. By the result quoted above, if μ is a normal measure on X with support equal to X, there is a measure equivalent to μ which is T-invariant if and only if there are no wandering measurable subsets E of X.

If such a set E did exist, such that the sets $\{T^{nk}(E)\}$ are disjoint, for some infinite sequence of integers nk, then by regularity of μ we can assume that E is closed. Since X is a stonean space, E is both open and closed ('clopen'). Thus the characteristic function χ_E corresponds to a projection in the algebra R and the set of projections $\theta^{nk}(\chi_E)$ is an orthogonal set. From the algebraic perspective then we can say the following. Given an abelian von Neumann algebra R, an automorphism θ of R and a faithful normal state acting on R. Then there is a faithful normal θ-invariant state acting on R if and only if there are no non-trivial projections E in R such that for some infinite sequence of integers nk, the projections $\theta^{nk}(E)$ are mutually orthogonal. It can be easily shown that for a commutative algebra this condition on the set of projections is equivalent to the requirement that there are no nonzero projections E with $\theta^{nk}(E) \to 0$ in the ultraweak topology as $nk \to \infty$ for some infinite sequence nk of integers. This new formulation now generalizes easily to the noncommutative (quantum) case as follows. Let R be a separable von Neumann algebra, G a group of automorphisms of R. Then a nontrivial projection E in R is *wandering* if E is such that; $g_{nk}(E) \to 0$ for some infinite sequence g_{nk} in G. Convergence is defined in the weak operator topology.

If $A(x)$ is a fibre algebra then it is a von Neumann algebra with trivial centre and is countably decomposable. Let $\alpha : g \to \alpha_g$ be a group representation of the translation subgroup T of the Poincare group which is ultraweakly continuous.

There is a faithful normal diffeomorphism- invariant quantum state on the fibre algebra $A(x)$ if and only if there are no wandering projections in $A(x)$.

Proof. if E is a projection in $A(x)$ such that $\alpha_{g_{nk}}(E) \to 0$ for some infinite sequence g_{nk} in G and f is a faithful, normal α-invariant state, then $f(E) = 0$, thus $E=0$.

The proof of the converse is based on work by M Takesaki on singular states [18].

We assume that there are no wandering projections in $\mathbf{A(x)}$. The fibre algebra $\mathbf{A(x)}$ has a faithful normal state f. By the HK - fixed point property (see earlier), applied to the set; $\mathcal{E}$ =weak* closed convex hull of $\{v(g)f; g \in T\}$, $\mathbf{A(x)}$ has an invariant state which we denote as h. We need to show that h is both normal and faithful. By [18] h has a unique decomposition $h = h_n + h_s$ with h_n a normal positive linear functional and h_s a singular positive linear functional. By uniqueness of the decomposition, both of these linear functionals are also α-invariant. Let S be the support of h_n so that $0 \leq S \leq I$. If $S \neq I$ we can choose a projection F with $0 < F < I - E$ and $h_s(F) = 0$.

Let $\lambda = \inf_{g \in T}(f \circ \alpha_g(F))$. Since $h = h_n + h_s$, we have $h(F) = 0$. Therefore $\lambda = 0$. Thus there is a sequence g_{nk} with $f \circ \alpha_{g_{nk}}(F) \to 0$. Since f is faithful and normal this implies that $\alpha_{g_{nk}}(F) \to 0$ in the weak operator topology; i.e. F is a wandering projection.

This contradiction shows that the support of h_n equals I and h_n is the required normal, faithful diffeomorphism-invariant gravity state.

The Information dynamics of Black Holes

In thermodynamics, entropy is defined through considering the phase space of a dynamical system. The emergent behaviour of a classical system gives rise to regions of phase space, each corresponding to similar macro- level behaviour. The number and variation in size of these regions reflects the overall complexity of the system. This identification is known as 'coarse graining' of the phase space. The entropy of such a coarse grained region is essentially a count of all of the different micro-configurations constituting that region; i.e. it is a measure of the volume V of that region in phrase space. Boltzmann's formula for entropy is given by $k \, logV$ where k is Boltzmann's constant with the logarithm ensuring that entropy is additive across phase space configurations. The structure of the phase space is such that each set of initial conditions (x,p) generates a unique solution $S(x,p)$.

A solution starting in a low entropy state will tend to wander into larger coarse grained volumes. Hence thermodynamic entropy tends to increase over time if the system is isolated;

giving rise to the second law of thermodynamics. We can see that the law is more valid in a statistical sense when applied to an ensemble of identical systems; the expected entropy of the ensemble will drift to higher values over time.

J von Neumann defined a quantum version of entropy as follows: Let f be a normal state of a local algebra of observables $O(D)$ acting on the Hilbert space H. Then for any operator A in $O(D)$ we have

$$f(A) = \sum_j \langle x_j, Ax_j \rangle \text{ where } \{x_j; j=1,2,3,...\} \text{ form an orthogonal set with } \sum_j \| x_j \|^2 = 1.$$

We can write this as $f = \sum_j p_j^2 \omega_{y_j}$ where

ω_{y_j} is a vector state with $\| y_j \| = 1$, and $0 \le \sum_j p_j^2 = \sum \| x_j \|^2 = 1.$

Thus f is a convex sum of pure states.

Von Neumann's non-commutative equivalent of a partition is the density operator

$\rho = \sum_k p_k^2 y_k \rangle\langle y_k$ ie the weighted sum of the projections E_k onto the vector spaces

$E_k H = [| y_k \rangle]$ Then we have the well-known equivalence;

$$f(A) = \sum_j p_j^2 \langle y_j, Ay_j \rangle = \sum_{j,m,n} p_j^2 < y_j, m >< m | A | n >< n, y_j > \text{ with } \{m,n\} \text{ a discrete basis;}$$

$$= \sum_{j,m,n} < n, p_j^2 y_j >< y_j, m >< m | A | n > = \sum_{m,n} \langle n | \rho | m \rangle < m | A | n > = \sum_{j,n} \langle n | \rho A | n \rangle$$

$$= Trace(\rho A)$$

For such a normal state f, the von Neumann entropy is defined as $-\sum_j p_j^2 \log p_j^2$. We interpret it as an (inverse) measure of the amount of information that the quantum system in a given state will yield through measurement. The larger the entropy of the quantum system, the less information can be extracted.

The von Neumann Entropy of a Black Hole

The measurement process cannot be performed by an external observer to elements within the interior, beyond the event horizon. We thus partition the event horizon of the black hole with elements each of area l_P^2, where l_P is the Planck length ($\cong 1.6 \times 10^{-35} m$) and assume the Planck area l_P^2 corresponds classically to the minimal projection $E_k H = [| y_k \rangle]$ of the pure vector state

68

ω_{y_k}. Let N be the total finite number of partitions. By the 'no hair' hypothesis there is no preferred location on the event horizon, so that each partition element must have the same weighting. The von Neumann entropy of this partition is thus given by;

$$-\sum_1^N \frac{1}{N}\log\left(\frac{1}{N}\right) = \log N = \log\frac{S}{l_p^2} = \log\frac{c^3 S}{\hbar G}$$ with S the surface area of the black hole, which is

similar to the Beckenstein-Hawking formula.

This relates to classical relativity via the macro- level Clausius definition of entropy E;

$$\oint \frac{\delta Q}{T} = \Delta E$$ where, integrated over a Carnot cycle, a change in energy δQ results in a change in

entropy corresponding to a change in space-time curvature.

1. R Wald (1994) Quantum Field Theory in Curved Space-Time and Black Hole Thermodynamics, Univ of Chicago Press.

2. Pontryagin L (1966) Topological groups 2nd Edn. Gordon & Breach.

3. Moffat J, Oniga T, and Wang C. (2017) Unitary representations of the translational group acting as local diffeomorphisms of space-time. J Phys Math 8:2 DOI: 10.4172/2090-0902.1000233

4. Douglas R (1972) Banach algebra techniques in operator theory. Academic Press.

5. Moffat J (1977) On groups of automorphisms of operator algebras. Math Proc Camb Phil Soc 81: 237-241

6. Hajian, A. and Kakutani, S., (1964). Weakly wandering sets and invariant measures. *Transactions of the American Mathematical Society*, *110*(1), pp.136-151.

7. Moffat J (1974) Automorphisms of Operator Algebras, PhD Thesis Univ. of Newcastle on Tyne, UK.

8. Moffat J, Wang C (2017) Factorial Unitary Representations of the Translational Group, Invariant Pure States and the Supersymmetric Graviton. J Phys Math 8: 257. Doi: 10.4172/2090-0902.1000257

9. Bailin D, Love A (1994) Supersymmetric gauge field theory and string theory. Taylor and Francis

10. Kadison R, Ringrose J (1983) Fundamentals of the theory of operator algebras. Academic Press.

11. Moffat J91974) 'On Groups of Automorphisms of the Tensor Product of von Neumann Algebras' Math Scand 34.

12. Jacobson T (1995) 'Thermodynamics of Spacetime: The Einstein Equation of State' Phys. Rev. Lett. 75.

13. Moffat J (1973) 'Continuity of Automorphic Representations' Math Proc Camb Phil Soc 74.

14. Dunford N, Schwartz J (1958) Linear operators. Vol I, Interscience, New York.

15. Ryll-Nardzewski C (1967) 'On Fixed Points of Semi-Groups of Endomorphisms of Linear Spaces'. Proc. 5th Berkeley Symp. Probab. Math. Stat. Univ. California Press 2(1).

16. Namioka I, Asplund E (1967) 'A Geometric Proof of Ryll-Nardzewski's Fixed Point Theorem.' Bull. Amer. Math. Soc. 73.

17. Dixmier, J (1951)1 Sur certains espaces considérés par MH Stone. Instituto Brasileiro de Educação.

18. Takesaki, M (1959) On the singularity of a positive linear functional on operator algebra. Proceedings of the Japan Academy 35.7.

19. Stormer, E (1973) Automorphisms and equivalence in von Neumann algebras Pacific Journal of Mathematics 44.1.

20. Hopf, E (1932) Theory of measure and invariant integrals." Transactions of the American Mathematical Society 34.2.

21. Halmos, P (1956) Lectures on ergodic theory. Vol. 142. American Mathematical Soc.

22. Reed, M, and B Simon (1980) Methods of modern mathematical physics. vol. 1. Functional analysis. Academic Press.

23. Pier, J (1984) Amenable locally compact groups. Wiley-Interscience.

24. Segal, I (1951) A class of operator algebras which are determined by groups.Duke Mathematical Journal 18.1.

25. Krein, M, and D Milman (1940) On extreme points of regular convex sets."Studia Mathematica 9.1..

Additional Bibliography

Sakai, S (2012) C*-algebras and W*-algebras. Springer Science & Business Media,.

McMurray S (1993) 'Quantum Mechanics' Addison Wesley..

Sakurai J (1994)'Modern Quantum Mechanics; Revised Edition', Addison Wesley.